Framework FOCUS

Science

Dictionary

11-14

Bob McDuell

Published by Letts Education
The Chiswick Centre
414 Chiswick High Road
London W4 5TF
tel: 020 89963333
fax: 020 87428390
email: mail@lettsed.co.uk
website: www.letts-education.co....

Letts Educational is part of the Granada Learning Group. Granada Learning
is a division of Granada plc.

© Bob McDuell 2003

First published 2003

ISBN 1 84085 457 X

The author asserts the moral right to be identified as the
author of this work.

British Library Cataloguing in Publication Data

A catalogue record for this book is available from the British Library.

Acknowledgements
Every effort has been made to contact the holders of copyright material,
but if any have been inadvertently overlooked the publishers will be
pleased to make the necessary arrangements at the first opportunity.

The publishers would like to thank the following for permission to
reproduce photographs and company logos (T = Top, B = Bottom,
C = Centre, L= Left, R = Right):
Heather Angel, 20, 38, 64; Martyn Chillmaid, 8, 67, 83; Mark Fuller/Leslie
Garland Picture Library, 72; Paul Mulcahy, 87; S Summerhays/Natural
Visions, 99; NHPA/S Dalton, 4CL, 4CR, G I Bernard, 4T, J Warwick, 13, G
Edwards, 40; Science Photo Library/Simon Fraser, 9, Charles D Winters, 23.

Cover photograph: Dr Jeremy Burgess/Science Photo Library

Illustrations
Ken Vail Graphic Design, Cambridge

Commissioned by Helen Clark

Project management by Vicky Butt

Editing by Nancy Candlin

Design by Ken Vail Graphic Design, Cambridge

Production by PDQ

Printed and bound by Canale, Italy

How to use this dictionary

The Letts Science Dictionary is aimed at Key Stage 3 Science students, although other students will also benefit from its clear explanations and interesting detail. Text and layout have been designed to make the dictionary easy to use, including many features to help you understand as much about the words as about the scientific ideas. This will make you more confident when you meet the words in your reading, and when you use them in your writing. These features are described in the examples below.

Entry word, or headword
The main form of the word. Unusual forms of the word are given in brackets after the headword.

Definition
The meaning of the word. This is kept as clear and concise as possible.

Information
Background information on the word and its use.

Symbol
Any symbol or abbreviation of the headword is given in square brackets.

Example
An example of the headword being used in a sentence.

Related words
Other entries in the dictionary related to the headword.

Topic area
The area in Science in which the word is most commonly used (see full list on the next page).

Part of speech
This tells you the job that the word does in a sentence.

Other meanings
If a word has more than one meaning, each meaning is numbered and each is followed by its own example.

friction

noun
Friction is the force that opposes the relative motion of two bodies in contact.

eg Friction is reduced by using a lubricant such as oil. Friction is important for brake linings, soles of shoes and tyres

i The word 'friction' comes from the 16th-century Latin word *fricare*, which means 'to rub'.
➡ lubricant

ampere [A]

noun
The ampere is the SI unit of electric current. Current is measured on an ammeter.

eg The current flowing through a circuit was one ampere.
➡ ammeter

counterbalance

1 noun
A counterbalance is a weight or force that balances or cancels out another force.

eg In the diagram, the small weight is acting as a counterbalance to the large one.

2 verb
To counterbalance means to balance out a weight or force.

Pronunciation
How to say the word.

hydraulic (*high-**draw**-lick*)
adjective
Something is said to be hydraulic if it is operated by pressure transmitted through a pipe by a liquid.

eg The diagram shows a simplified representation of the hydraulic braking system in a car.

Illustration
Visual examples rather than verbal examples are given where appropriate.

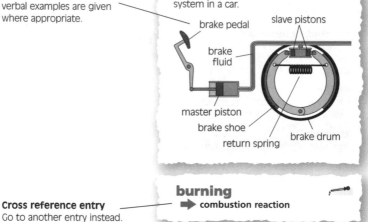

brake pedal — slave pistons

brake fluid

master piston
brake shoe
return spring — brake drum

burning
➡ **combustion reaction**

Cross reference entry
Go to another entry instead.

Topic areas

Each entry is labelled with an icon which tells you its topic area:

? ? ?	Scientific enquiry
🐝	Biology
🔥	Chemistry
⊏⊗	Physics

abdomen

noun

In animals with a backbone, the abdomen is the area below the thorax that contains the digestive organs. In mammals it is separated from the thorax by the diaphragm.

eg The abdomen contains the stomach, intestines, liver and kidneys.

➡ **diaphragm, thorax**

absorption

noun

Absorption is the process by which digested particles of food pass from the digestive system into the bloodstream.

eg Glucose formed in the stomach is absorbed into the bloodstream. This process is called absorption.

acceleration

(*ack-**sell**-er-a-tion*)

noun

The rate of change of the velocity of a moving body is called acceleration. It is measured in units of m/s^2. Acceleration will only take place when a body is acted upon by an unbalanced force. A negative value for acceleration shows that an object is slowing down (decelerating).

eg An object speeds up from 10 m/s to 30 m/s in 5 seconds. The average acceleration is $\frac{30-10}{5} = 4$ m/s^2.

Calculate the average acceleration when a stationary object speeds up to 25 m/s in 5 seconds.

➡ **force**

accurate

? ? ?

adjective

Measurements that are accurate are exact or correct.

eg Measuring the length of a piece of paper with a ruler can give accurate results.

acid

noun

An acid is a compound containing hydrogen that, when dissolved in water, forms a solution with a pH value of less than 7.

eg Common acids are sulphuric acid H_2SO_4, hydrochloric acid HCl, nitric acid HNO_3 and ethanoic acid (vinegar). Which chemical element is in all acids?

🔢 The word 'acid' comes from the Latin word *acidus*, which means 'sour'.

acid rain

noun

Rain that has a pH value of less than 5 is called acid rain. It is caused by dissolved sulphur dioxide and oxides of nitrogen forming when fossil fuels are burned. Acid rain causes the erosion of rocks such as limestone, and the corrosion of metals. It is also linked to the damage and death of trees in forests and the acidification of lakes.

eg Acid rain can be reduced by removing sulphur dioxide from power station chimneys and by using catalytic converters to remove nitrogen oxides from car exhausts.

addiction

noun

Addiction is the state of dependence caused by continually using drugs, alcohol or other substances. When these substances are not available, withdrawal symptoms are shown. Repeated use of the

A B C D E F G H I J K L M N O P Q R S T U V W X Y Z

substances can alter body chemistry. Addiction is treated by gradually reducing the dose.

eg Drugs such as heroin can result in addiction.

adolescence
noun

Adolescence is the stage of development between puberty and adulthood.

eg A lot of physical and emotional changes can occur during adolescence.

ℹ The adjective 'adolescent' and the noun 'adolescence' come from the Latin word *adolescere*, which means 'to grow up'.

aerobic
adjective

Respiration is said to be aerobic when it describes how living organisms use oxygen for the efficient release of energy. Most organisms, except certain bacteria, yeasts and internal parasites, respire aerobically.

eg Aerobic respiration can be summarised by the equation:

$$C_6H_{12}O_6 + 6O_2 \rightarrow 6CO_2 + 6H_2O + 2880kJ$$

ℹ The word 'aerobic' comes from the two Greek words, *aero* and *bios*, for 'air' and 'life'.
➡ **anaerobic**

air
noun

Air is the mixture of gases that surrounds the Earth. It consists of approximately 80% nitrogen and 20% oxygen, with tiny amounts of water vapour, carbon dioxide, argon and other gases.

eg The composition of air can vary from place to place. In a city there is likely to be less oxygen and more carbon dioxide and other gases such as oxides of nitrogen and sulphur dioxide, than in the country.

The composition of air

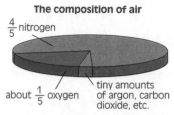

$\frac{4}{5}$ nitrogen

about $\frac{1}{5}$ oxygen

tiny amounts of argon, carbon dioxide, etc.

➡ **atmosphere**

air resistance
noun

Air resistance is a force that slows down an object as it moves through the air.

eg Cars are tested in a wind tunnel to check their air resistance. The lower the air resistance, the less fuel a car will use.

albumen
noun

Albumen is the white of an egg.

eg The albumen contains proteins and provides food for the hatching chick.

alcohol
noun

An alcohol is one of a family of compounds containing a hydroxyl (or –OH) group. Ethanol is the most common alcohol and is sometimes referred to simply as alcohol. Ethanol has a formula C_2H_5OH.

eg The alcohol called ethanol is made from the fermentation of starch or sugar solutions with yeast.

ℹ The word 'alcohol' comes from the Arabic words *al* and *kuhl*.

align (*a-line*) ???
verb

To align means to bring things into line.

eg When collecting results, we attempt to align them.

ℹ The verb 'align' and the noun 'alignment' come from the French *à ligne*, which means 'into line'.

alkali

noun

An alkali is a compound, usually a hydroxide that, when dissolved in water, forms a solution with a pH value greater than 7.

eg One common alkali is sodium hydroxide [NaOH].

The word 'alkali' comes from the 14th-century Arabic word *alqali*, which means 'calcinated ashes' and refers to the alkali produced when lime is burned.

alloy

noun

An alloy is a mixture of metals or a mixture of metal and carbon.

eg Brass is an alloy of copper and zinc.
➡ **steel**

alveolus (plural: alveoli)

noun

An alveolus is one of many thousands of tiny air sacs in the lungs where gaseous exchange takes place.

eg More than one alveolus are called alveoli. The diagram shows some alveoli and demonstrates how gaseous exchange takes place.

Alveoli

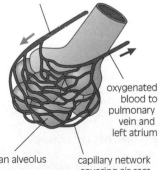

de-oxygenated blood from pulmonary artery and right ventricle

oxygenated blood to pulmonary vein and left atrium

an alveolus

capillary network covering air sacs

Inside an alveolus

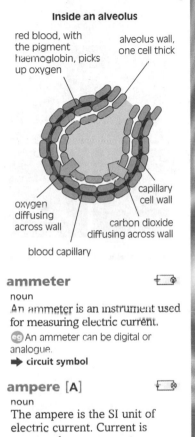

red blood, with the pigment haemoglobin, picks up oxygen

alveolus wall, one cell thick

capillary cell wall

oxygen diffusing across wall

carbon dioxide diffusing across wall

blood capillary

ammeter

noun

An ammeter is an instrument used for measuring electric current.

eg An ammeter can be digital or analogue.
➡ **circuit symbol**

ampere [A]

noun

The ampere is the SI unit of electric current. Current is measured on an ammeter.

eg The current flowing through a circuit was one ampere.

The ampere is named after the 19th-century French physicist, André Marie Ampère.
➡ **ammeter**

amphibian (*am-fi-bee-an*)

noun

An amphibian is an animal with a backbone that spends its larval stage in water, moving to the land when it matures and returning to water to breed.

eg A frog is an amphibian, as is a toad. They lay eggs in water and their tadpoles also live in water, but the mature frogs and toads come out of the water.

amplitude

noun

The maximum displacement of a wave from its rest position is called the amplitude.

eg If the amplitude of a sound wave is increased, the sound becomes louder.

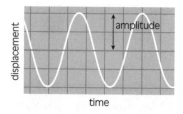

anaerobic

adjective

Respiration is said to be anaerobic when it takes place without oxygen. An athlete in a long distance race may start to undergo anaerobic respiration, which leads to the build up of lactic acid and, possibly, cramp. The word equation for anaerobic respiration is:

glucose → lactic acid + energy

eg Fermentation is an example of anaerobic respiration.

➡ aerobic

analyse

verb ? ? ?

To analyse something is to investigate or examine it by breaking it down into parts.

eg Sue is going to analyse the powder to find out the chemicals it contains.

analysis

noun ? ? ?

Analysis is the examination of something by breaking it down and identifying its separate parts.

eg Analysis of mineral water is carried out before it is bottled and its composition is printed on the label.

➡ synthesis

anomalous result

noun ? ? ?

In an experiment, an anomalous result is one that does not fit a trend or a pattern. If a graph is plotted, an anomalous point does not fit on the line of best fit. When you evaluate an experiment, you should recognise and try to explain any anomalous results.

eg The diagram shows an anomalous result.

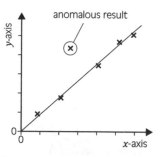

The word 'anomalous' comes from the Greek words *an-* and *homalos*, which mean 'not even'.

antagonistic muscle

noun

One of a pair of muscles that work together in a joint in the skeleton is called an antagonistic muscle. For example, when the arm is extended, one set of muscles relaxes while the other set contracts.

🟢 The diagram shows antagonistic muscles in the arm. The biceps is the antagonistic muscle of the triceps and vice versa.

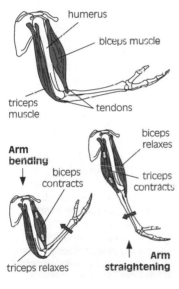

humerus

biceps muscle

triceps muscle

tendons

Arm bending

biceps contracts

triceps relaxes

biceps relaxes

triceps contracts

Arm straightening

antibiotic

noun

An antibiotic is a substance that can selectively destroy or inhibit bacteria or fungi.

🟢 An antibiotic can be used to treat bacterial infections but not infections caused by viruses.

🔳 Sir Alexander Fleming found the first antibiotic by accident in 1928. It was penicillin. It took 15 years to produce penicillin in a form that could be used as a medicine.

antibody

noun

An antibody is a protein produced by white blood cells as part of the body's defence system. These chemicals attack bacteria or mark them so they can be destroyed by white blood cells.

🟢 When you have influenza, your body quickly produces lots of the antibody that attack and destroy the bacteria that cause the illness.

antigen

noun

An antigen is a substance that causes the production of antibodies.

🟢 One such antigen is the proteins on the surface of bacteria, viruses and pollen grains.

apparatus

plural noun **? ? ?**
Equipment used in the laboratory is called apparatus.

🟢 Beakers, flasks, thermometers and measuring cylinders are all pieces of apparatus.

approximate   **? ? ?**

adjective

Something is said to be approximate when it is nearly correct, but not exact.

🟢 How could you measure out an approximate volume of water?

area   **? ? ?**

noun

The measurement of the surface of a shape or object is its area.

🟢 An investigation showed that the area of a leaf was $10\,cm^2$.

➡ surface area

artery

noun

An artery is a blood vessel that carries blood away from the heart. Because the blood is under high pressure, the walls of arteries have to be thick.

🟢 One artery takes blood to the lungs; another takes it to the body.

➡ capillary, vein

A B C D E F G H I J K L M N O P Q R S T U V W X Y Z

5

asexual reproduction

noun

Asexual reproduction occurs when cells divide to produce identical copies of themselves. All offspring from asexual reproduction have identical genes to their parents. These offspring are called clones.

eg Taking cuttings of an existing plant is an example of asexual reproduction, and produces plants that are identical to the original plant.
➡ clone, mitosis

ash

noun

Ash is the powdery residue produced when something has been burned.

eg The ash left after a bonfire is alkaline and is a good source of potassium for plants.

association (verb: associate)

noun

Association is one of the functions of the central nervous system. Information from sense organs is collected and processed. This information is then used to cause other activities within the body.

eg When you smell dinner cooking, the association your central nervous system makes between the smell and eating causes saliva to be produced in your mouth.

asteroid

noun

An asteroid is a fragment of rock that orbits around the Sun, mainly between the orbits of Mars and Jupiter.

eg Some people believe that dinosaurs were wiped out on the Earth when the Earth was hit by an asteroid.

🔝 The word 'asteroid' comes from the Greek word *asteroides*, which means 'star-like'.

-ate

suffix

A salt that contains oxygen has a name with the suffix '-ate'.

eg Sodium sulphate is a salt containing sodium, sulphur and oxygen.

atmosphere

noun

The air around the Earth is called the atmosphere. It is prevented from escaping by the planet's gravitational attraction.

eg Without the Earth's atmosphere, life on Earth would be impossible.
➡ air

atom

noun

An atom is the smallest part of matter that can exist and take part in a chemical reaction. Atoms are too small to be seen with a microscope. The largest atom, caesium, has a diameter of 0.0000005 mm.

eg A piece of iron is made up of iron atoms and a piece of copper is made up of copper atoms. Iron and copper atoms differ in size and mass.

🔝 About 2400 years ago, the Greek philosopher Democritus suggested that all matter was made up of atoms. He had no experimental evidence for this and the idea was not generally accepted until about 200 years ago. John Dalton proposed that matter was made of atoms but he believed that atoms could not be split up.

at rest

adjective

Something is said to be at rest if it is not moving.

eg After rolling along the floor, the marble stopped and was at rest.

attraction (verb: attract)

noun

Attraction is a force that tends to pull two objects closer together.

🔵 There is attraction between the north pole of one magnet and the south pole of another.

➡ repulsion

average　　　? ? ?

noun

A number that is representative of a group of numbers is called an average. The average of n numbers is found by adding them together and dividing by n.

🔵 The average of 17, 23 and 11 is found by adding them and dividing by 3. What is the average of these three numbers?

📕 The word 'average' comes from the Arabic word *awariya*, which means 'damaged goods'.

axis (plural: axes)　　? ? ?

1 noun

The imaginary straight line around which a planet rotates is called the axis.

🔵 The diagram shows the Earth and the axis around which it rotates.

Earth's axis

2 noun

An axis is one of the lines of reference used on a graph to specify a position.

🔵 When drawing a graph, we use a horizontal axis called the x-axis and a vertical axis called the y-axis.

Bb

baby
noun

A baby is a newborn or very young child or animal.

eg A human baby needs regular attention from its parents.

fi The word 'baby' dates back to about the 13th century and probably imitates the sound a human baby makes.

bacteria (singular: bacterium)
noun

Bacteria are a very large group of micro-organisms, including some that produce infectious diseases. Bacteria are destroyed by antibiotics.

eg The diagram shows a bacterium.

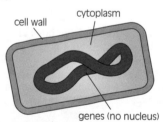

cell wall cytoplasm
genes (no nucleus)

fi The word 'bacteria' comes from the Greek word *bakterion*, which means 'little stick', and this describes the shape of some bacteria.
➡ antibiotic, virus

baking powder
noun

Baking powder is used to get a cake to rise when baking. The main ingredients of baking powder are sodium hydrogencarbonate (bicarbonate of soda) and tartaric acid. When added to flour and water and heated, the mixture starts to rise.

Bubbles of carbon dioxide are formed and these make the cake rise.

eg Which cooking flour already contains baking powder?

balance
1 noun ???

A balance is an instrument used for weighing.

eg The photograph shows a balance that can weigh to the nearest 0.01 g.

2 noun

Balance is a state in which two opposite forces are exactly equal.

eg When a skydiver is falling at constant velocity, there is a balance between the upward and downward forces.

air resistance
(upward force)

constant velocity

gravity
(downward force)

balanced diet

noun

A balanced diet contains the range of nutrients needed to stay healthy.

eg For a balanced diet you must eat proteins, carbohydrates, fats, vitamins and minerals.

bar chart ???

noun

A bar chart is a chart which displays values by means of vertical or horizontal bars. These bars are separated by gaps.

eg Two types of bar chart are shown below.

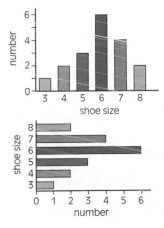

bar line graph ???

noun

A bar line graph is similar to a bar chart but the bars are replaced by lines.

eg This bar line graph shows the results of a shoe-size survey in class 7c.

basalt (*ba-salt*)

noun

Basalt is an igneous rock formed when magma crystallises rapidly on the surface of the Earth. It usually consists of small black or dark-coloured crystals.

eg Basalt is the most common volcanic rock.

bath salts

noun

Bath salts contain sodium hydrogencarbonate and sodium carbonate. This softens hard water (containing dissolved calcium and magnesium compounds) so soap lathers well.

eg Colouring and a pleasant perfume are usually added to bath salts to make them more attractive.

beaker

noun

A beaker is a piece of apparatus, usually made of glass, used to hold liquids. It has a lip to make pouring easier.

eg The diagram shows a beaker in three dimensions and in cross-section.

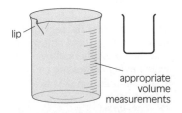

lip

appropriate volume measurements

A B C D E F G H I J K L M N O P Q R S T U V W X Y Z

bi-
prefix

The prefix 'bi-' means 'two'.

eg A bimetallic strip is a strip made of two metals fixed together.

➡ di-

bicarbonate of soda
noun

Bicarbonate of soda is a weak alkali used in bath salts, baking powder, indigestion remedies and some toothpastes. The chemical name is sodium hydrogencarbonate and the chemical formula is $NaHCO_3$.

eg Bicarbonate of soda can be used to clear up acid spills in the laboratory.

biomass
noun

The total mass of living organisms in an ecosystem, population or designated area at a particular time is called the biomass.

eg Pyramids of biomass show how much food is available at each feeding level.

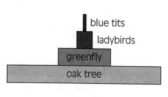

blue tits
ladybirds
greenfly
oak tree

biosphere
noun

The biosphere is the region of the Earth's surface (land and water) and the atmosphere above it in which living organisms can live.

eg The biosphere can also be called the ecosphere.

➡ water cycle

boiling temperature
noun

The boiling temperature (sometimes called boiling point) is the temperature at which a liquid changes to a gas.

eg The boiling temperature of water is 100°C.

➡ change of state, evaporation

botany
noun

Botany is the branch of Biology concerned with the study of plants.

eg A scientist who studies botany is called a botanist.

brain
noun

The brain is the soft tissue inside the skull of vertebrates that controls thoughts, memories and emotions.

eg Dolphins are known to have a large brain.

breathing
noun

Breathing is the process by which air is drawn into and then expelled from the lungs. When breathing in (called inhaling):
● the intercostal muscles move the ribs up and out
● the chest cavity becomes larger so air pressure is reduced
● air rushes in to fill the lungs.
When breathing out (called exhaling) the opposite happens.

eg Breathing is a mechanical process that gets air into the lungs and removes waste, including carbon dioxide.

Inhalation

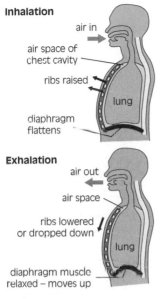

air in

air space of chest cavity

ribs raised

lung

diaphragm flattens

Exhalation

air out

air space

ribs lowered or dropped down

lung

diaphragm muscle relaxed – moves up

➡ **alveolus, bronchiole, bronchus, diaphragm, exhale, inhale, lung, trachea**

breed

verb

To breed means to reproduce sexually.

🐞 When plants or animals breed sexually, the offspring are not identical to the parents.

brittle

adjective

Something is said to be brittle if it breaks easily when bent or hit.

🐞 Sulphur is a brittle solid.

🔧 The word 'brittle' comes from the Anglo-Saxon word *breotan*, which means 'to break in pieces'.

bronchiole (*bron-ki-ole*)

noun

A bronchiole is a tube in the lungs carrying air from the wider bronchus to thousands of air sacs or alveoli.

🐞 Bronchioles can be damaged by infection and smoking.

➡ **alveolus, bronchus, lung, respiratory system, trachea**

bronchus (plural: bronchi) (*bron-kus*)

noun

A bronchus is one of a pair of large tubes running from the trachea to the lungs.

🐞 The bronchus is surrounded by rings of cartilage to make it rigid and prevent it collapsing when there are changes in pressure.

➡ **alveolus, bronchiole, lung, respiratory system, trachea**

bubbles

noun

Bubbles are formed when a gas is released during a chemical reaction. They are small amounts of gas trapped in a liquid.

🐞 When an acid is added to a carbonate, bubbles of carbon dioxide are formed.

burette (*byou-ret*)

noun

A burette is a piece of apparatus capable of measuring the volume of a liquid to the nearest $0.05 \ cm^3$.

🐞 The diagram shows a burette.

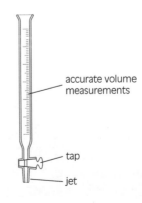

accurate volume measurements

tap

jet

burning
➡ **combustion reaction**

buzzer

noun

A buzzer is an electrical device that makes a buzzing sound, often used as a warning.

eg A deep freeze may have a buzzer fitted so that if the electricity supply fails there is a warning before the food thaws.

Cc

calculate (noun: calculation)

verb

To calculate is to use numbers to work out values.

eg Calculate the average value of three measurements.

capillary (plural: capillaries)

noun

A capillary is the narrowest type of blood vessel in the human body.

eg A capillary is a small blood vessel with thin walls. Capillaries link veins and arteries together.

➡ artery, vein

carbohydrate

noun

A carbohydrate is a substance that produces energy in respiration. It contains three elements: carbon, hydrogen and oxygen. There are always twice as many hydrogen atoms as oxygen atoms in a carbohydrate molecule.

eg Glucose, sucrose and starch are carbohydrates. Why is fructose, $C_6H_{12}O_6$, a carbohydrate but alcohol, C_2H_5OH, not?

carbon

noun

Carbon is a chemical non-metallic element.

eg Carbon exists in two forms – diamond and graphite.

carbonate

noun

A salt of carbonic acid, H_2CO_3, is called a carbonate.

eg Chalk, limestone and marble are three forms of calcium carbonate.

carbon dioxide [CO_2]

noun

Carbon dioxide is a colourless, odourless gas produced in many reactions, including when carbon or carbon compounds burn, during respiration, during fermentation and from reactions of carbonates. The test for carbon dioxide is to bubble it through colourless limewater. Carbon dioxide turns the limewater white (like milk).

eg Carbon dioxide is produced when carbonates are heated and when carbonates react with acids.

carnivore

noun

A carnivore is an animal that eats flesh.

eg A tiger is a carnivore.

carry out a survey ? ? ?

verb phrase

To carry out a survey means to collect information about something in order to learn more about it.

eg The children were asked to carry out a survey on the number of creatures living in the school pond.

catalyst (*cat-a-list*)

noun

A catalyst is a substance that alters the rate of a chemical reaction but is not itself used up.

eg Hydrogen peroxide decomposes into water and oxygen. Manganese (IV) oxide is a catalyst for this reaction.

The word 'catalyst' comes from the Greek word *katalysis*, meaning 'breaking up'.

cathode ray oscilloscope [CRO]

➡ oscilloscope

caustic (*core-stick*)

adjective

A substance is said to be caustic if it is strongly alkaline and corrosive to human tissue.

eg Alkalis, such as sodium hydroxide, are caustic. Special precautions must be taken when using caustic substances.

CD-ROM ? ? ?

noun

CD-ROM (Compact Disk Read-Only Memory) is a method of storing information on a disk that can be used in a computer.

eg Some textbooks are available with an accompanying student CD-ROM.

cell

1 noun

A cell is the smallest unit of an organism that can function. Cells are the building blocks of all living things.

eg The diagrams show a plant cell and an animal cell. What is present in both? What is present in a plant cell but not in an animal cell?

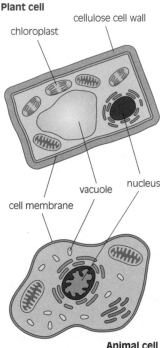

Plant cell

chloroplast

cellulose cell wall

vacuole

nucleus

cell membrane

Animal cell

2 noun

A cell is a device consisting of two metal rods (called electrodes) dipping into a solution that conducts electricity (the electrolyte). A cell converts chemical energy into electrical energy.

eg A cell is used to power an electrical circuit.

The word 'cell' comes from the Latin word *cella*, used to refer to a small room.

Celsius [C]

adjective

Celsius refers to the Celsius scale used for measuring temperature. It used to be called the Centigrade scale.

eg On the Celsius scale, the freezing point of water is 0°C and the boiling point of water is 100°C.

i The Celsius scale was first developed in 1742 and was named after the Swedish astronomer, Anders Celsius.

cement

noun

Cement is a powdery substance used in building. It is usually used as an ingredient in concrete or is mixed with water to make mortar. It is made by heating a finely-ground mixture of limestone and clay in a rotary kiln at 1500°C. When mixed with water it sets in a few hours.

eg The chemical composition of cement is complicated. It is best thought of as calcium aluminosilicate. Write down the names of the four elements that make up calcium aluminosilicate.

centi- ???

prefix

The prefix 'centi-' means 'one hundredth'.

eg A centilitre is one hundredth of a litre.

centimetre [cm] ???

noun

A centimetre is a unit of length equal to one hundredth of a metre.

eg A ruler is often 30 centimetres long.

➡ metre

change

verb

To change something is to make it different.

eg When magnesium is burned, the silvery metal turns to a white ash. This is a chemical change.

change of state

noun

A change of state refers to a change of the physical state of something, e.g. solid → liquid, liquid → gas, etc.

eg The diagram summarises the changes of state.

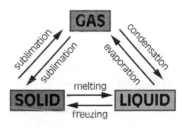

characteristic

noun

A distinctive quality or feature is called a characteristic.

eg Can you suggest a characteristic of an elephant?

charcoal

noun

Charcoal is an impure black form of carbon. It is formed when an organic material such as wood is heated out of contact with air.

eg Why is it important that wood is heated out of contact with air when charcoal is made?

chloride

noun

A salt of hydrochloric acid is called a chloride.

eg Common salt (sodium chloride, NaCl) is found in the earth as the mineral halite.

chlorine [Cl]

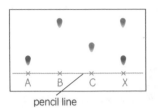

noun

Chlorine is a greenish-yellow gas in the family of halogens (group 7 of the Periodic Table). This element is used to kill micro-organisms in water before the water is used as tap water and in swimming pools. Chlorine is also used to make household bleaches.

(eg) Chlorine was used as a poison gas in World War 1.

chlorophyll (*claw-ro-fill*)

noun

Chlorophyll is the green pigment which gives green plants their colour. It is found within chloroplasts, mainly in the leaves of plants. It absorbs energy for use in the process of photosynthesis.

(eg) Chlorophyll absorbs the red and blue-violet parts of light but not green light, which is reflected. This gives plants their green colour.

🔎 The word 'chlorophyll' comes from the Greek words *chloro-* and *phyllon*, which mean 'green leaf'.

chloroplast

noun

A chloroplast is a tiny structure in the leaf or stem of a plant which holds chlorophyll. Most chloroplasts are close to the surface of a leaf where there is the maximum light intensity.

(eg) Chloroplasts are not found in animal cells or in root cells of plants.
➡ leaf

chromatogram

noun

A chromatogram is a piece of paper giving the results of a separation by chromatography.

(eg) The chromatogram shows that the dye D is a mixture of dyes A and B but not C. The spot for D produces two

spots – one at the same level as the spot for A and one at the same level as the spot for B. There is no spot at the same level as C.

![chromatogram diagram with spots labelled A, B, C, X and a pencil line]

pencil line

chromatography

noun

Chromatography is a technique used to separate a mixture of substances, usually in solution. Originally it was used to separate coloured mixtures such as plant pigments.
As the solvent moves up the piece of paper, the different substances in the mixture also move up the paper, but at different rates. This is because they have different solubilities in the solvent.
Each component forms a separate spot on the paper.

(eg) The diagram shows a chromatography investigation being carried out.

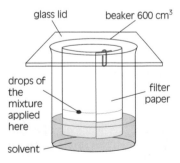

glass lid   beaker 600 cm³

drops of the mixture applied here

filter paper

solvent

🔎 The word 'chromatography' comes from two Greek words *chroma* and *graphein*, which mean 'colour writing'.

chromosome

noun

A chromosome is the part of a cell nucleus that carries the genes. Each chromosome consists of a long, coiled and folded strand of DNA.

eg The diagram shows the 23 pairs of chromosomes in a human male.

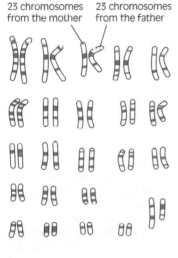

23 chromosomes from the mother   23 chromosomes from the father

➡ **gene**

cilia (singular: cilium) (*sill-y-a*)

noun

Cilia are small hair-like structures on the surface of certain cells. They beat rhythmically together to move the cell or to move the substances outside it.

eg In the air passages, cilia move mucus containing bacteria to the back of the throat. Here it is swallowed and the bacteria are destroyed in the acid conditions of the stomach.

circuit diagram

noun

A circuit diagram is a simplified diagram of an electric circuit. It uses standard circuit symbols. Connecting wires are shown by straight lines.

eg An example of a circuit diagram is illustrated below.

circuit symbols

noun

Scientists use a set of standard symbols called circuit symbols to represent common circuit components.

eg Here are some of the common circuit symbols.

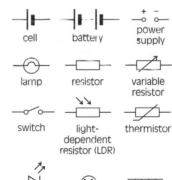

cell	battery	power supply
lamp	resistor	variable resistor
switch	light-dependent resistor (LDR)	thermistor
light-emitting diode (LED)	motor	heater
ammeter	voltmeter	

**A
B
C
D
E
F
G
H
I
J
K
L
M
N
O
P
Q
R
S
T
U
V
W
X
Y
Z**

circulation (verb: circulate)

noun

Circulation means the transport of blood around the body. In animals there is a system called the circulatory system, which transports blood around the body. The circulation of the blood in a human enables oxygen and glucose to be taken to the cells where respiration is taking place, and allows waste products to be taken away.

In mammalian circulation, blood is pumped from the heart to the lungs and then back to the heart. The blood is then pumped around the body through arteries, capillaries and veins. This is called double circulation.

➡ **artery, capillary, vein**

classification (verb: classify)

noun

Classification is an attempt to organise a large quantity of information into a logical order. Classifying living organisms into plants and animals is the first stage of classifying living organisms. Animals can then be divided into animals with backbones (vertebrates) and animals without backbones (invertebrates). Vertebrates can in turn be divided into cold-blooded (fish, amphibians, reptiles) and warm-blooded (birds and mammals). Invertebrates can also be subdivided into different groups. Chemical elements can be classified according to their chemical properties. The Periodic Table is a classification of the elements.

Classification enables organisation of knowledge and enables patterns to be seen.

clone (klone)

noun

A clone is a group of cells or organisms arising from asexual reproduction by a single parent individual. All clones have the same genetic information.

A clone produced by taking cuttings from a geranium plant will be identical in colour to other clones and to the parent plant.

➡ **asexual reproduction**

coil

noun

A coil of wire is an electrical wire wound into a spiral to make an electromagnet.

A coil will be found in an electric motor.

➡ **core, electromagnet**

collect

verb ? ? ?

To collect means to gather together.

The students were asked to collect examples of different rocks for an exhibition.

colour change

noun

A change in something is often accompanied by a change in colour and this is called a colour change. When reporting a colour change it is important to quote the colour both before and after.

The colour change in copper sulphate crystals when they are heated is from blue to white.

combustion reaction

noun

A combustion reaction is a chemical reaction that takes place during burning. The reaction involves a substance, called a fuel, catching alight and giving out energy in the form of heat and light. The products of a combustion reaction are oxides.

eg When carbon is heated in plenty of air, it catches alight and forms carbon dioxide. It gives out heat and light. This is a combustion reaction.

? The word 'combustion' comes from a 15th-century French word, *comburere*, which means 'to burn up'.

community

noun

In ecology, a community is all the organisms that live in a particular habitat.

eg A freshwater-pond community describes all the plants and animals that live in the pond.

compare

verb **? ? ?**

When you compare two things, you have to look at their similarities and differences. 'Compare' is one of the command words you might find in an examination question – it is telling you what to do.

eg Compare a plant cell and an animal cell. How do they differ? How are they the same?

? The word 'compare' comes from the Latin word *comparare*, which means 'to match'.

competition (verb: compete)

noun

Competition is the continual struggle for resources that organisms have with each other in order to survive.

eg A female cod will produce several million eggs in a single spawning. They cannot all survive as there are insufficient resources. There will be competition between the eggs for the resources that are available in order to survive.

complete circuit

noun

Electricity will only flow when there is a complete circuit with no gaps.

eg The diagrams show a complete circuit and an incomplete circuit.

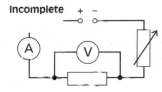

Complete

Incomplete

component

noun **? ? ?**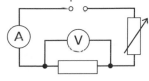

A component is any part of an instrument, engine, machine, etc.

eg A magnet is a component of an electric motor.

composition

noun

The parts from which something is made are its composition. A compound has a fixed composition.

eg The composition of silica glass is silicon dioxide, calcium carbonate and sodium carbonate

compound

noun

A compound is a chemical substance formed when two or more elements combine together in fixed proportions. A compound cannot be split up by physical means.

eg Water is a compound formed when 2 g of hydrogen and 16 g of oxygen are combined together.

A
B
C
D
E
F
G
H
I
J
K
L
M
N
O
P
Q
R
S
T
U
V
W
X
Y
Z

compressible

adjective

A gas is said to be compressible if, when the pressure is increased, the volume of the gas decreases. The gas particles are widely spaced and so when they are compressed they are forced closer together.

eg Solids and liquids are not compressible because the particles are already close together.

The word 'compressible' comes from the Latin word *comprimere*, which means 'to squeeze together'.

conclusion ? ? ?

noun

When you draw a conclusion, you make a judgement based on the evidence collected.

eg At the end of an investigation, you should try to use the evidence to draw a conclusion.

condense

(noun: condensation)

verb

To condense means to change from a gas to a liquid or a solid by cooling.

eg When cooled below 100°C, steam will condense to form liquid water.

condenser

noun

A condenser is used in a distillation apparatus to condense the vapour back to a liquid.

eg Cool water passes through an outer tube and condenses the vapour to a liquid. The diagram shows a condenser.

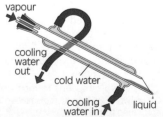

vapour

cooling water out

cold water

cooling water in

liquid

The condenser is sometimes called a Liebig condenser. Although the famous scientist Justus Liebig did not invent the condenser, he did make it popular.

condition ? ? ?

noun

The word 'condition' means the circumstance of something.

eg The best condition for storing milk is at low temperature in a refrigerator.

conduction

noun

Conduction is the way thermal or electrical energy is transferred in certain solids. When thermal energy is transferred, particles transfer their energy to neighbouring particles. When electrical energy is transferred, electrons pass through the solid.

eg Metals are good conduction materials for both thermal and electrical energy.

conductor

noun

A substance that conducts thermal or electrical energy is called a conductor.

eg Carbon is a good conductor of electrical energy but a poor conductor of thermal energy.

conglomerate

noun

Conglomerate is a sedimentary rock containing large fragments of rock.

eg Conglomerate is formed near river estuaries.

conifer

noun

A conifer is an evergreen shrub or plant with needle-like leaves. The leaves do not fall off in autumn. Conifers produce their pollen and seeds in cones.

eg A pine tree is a conifer.

The word 'conifer' comes from two Latin words, *ferre* and *conus*, which together mean 'to carry cones'.

conservation

noun

Conservation is the protection and preservation of the environment, its wildlife and its natural resources.

eg Zoos are able to assist conservation by ensuring that species of animals which are threatened in the wild can live and breed.

constant

noun ? ? ?

An unvarying number or quantity is called a constant.

eg The size of the Earth's pull is called its gravitational attraction. For the Earth, this constant has a value of 10 N/kg.

constant speed

noun

When an object is moving at a constant speed, it travels the same distance in each unit of time.

eg If the forces acting on a moving object are balanced, the object will move at a constant speed. If the forces acting on an object are unbalanced, the object will speed up or slow down.

consumer

noun

A consumer is any organism that cannot make its food but gets its food by consuming other organic material.

Herbivores are primary consumers as they eat vegetation. Carnivores are secondary consumers as they eat herbivores. Tertiary consumers eat other carnivores.

eg All animals are consumers, and plants such as saprophytic fungi, that live off decaying matter, are also consumers.

➡ carnivore, herbivore

control accuracy ? ? ?

verb phrase

To control accuracy is to choose the appropriate equipment when carrying out an investigation.

eg If you want to measure out small volumes of a liquid a number of times, a burette is the best piece of equipment to use to control accuracy.

control experiment

noun ? ? ?

A control experiment is an experiment in which the variable being tested in a second experiment is kept constant. This establishes the validity of the results of a second experiment.

eg In an investigation to find out what causes iron to rust, Jo set up a control experiment. In the control experiment, she let iron come into contact with air and water. In her other experiments, air was kept out in one case and water kept out in the other.

control variables

verb phrase ? ? ?

To control variables means to keep constant all but one of a set of variables that may alter the results of an investigation.

eg In planning an investigation, it is important to identify and control variables.

➡ fair test

A
B
C
D
E
F
G
H
I
J
K
L
M
N
O
P
Q
R
S
T
U
V
W
X
Y
Z

convection

noun

Convection is a way of transferring thermal energy through a liquid or a gas from a more dense region of lower temperature to a less dense region of higher temperature. It involves movement of particles.

eg The diagram shows how convection currents are set up in a room. Air at the top of the room is denser and colder. Particles from the colder air move downwards as particles from the hotter air move upwards.

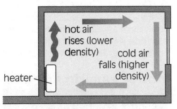

hot air rises (lower density)

cold air falls (higher density)

heater

convert

verb ??? 🦗 ⌇ 🔋

To convert is to change something from one form to another.

eg Boiling will convert water into steam.

cooling rate

noun

When a hot substance is left in colder conditions, the substance starts to cool.

The cooling rate is the drop in temperature divided by the time.

eg If water at 80°C cools to 50°C in 5 minutes, the average cooling rate is 6°C/minute. What is the cooling rate if water cools from 90°C to 30°C in six minutes?

core

1 noun

The central part of a planet such as the Earth is called the core.

eg The core of the Earth is shown in the diagram below.

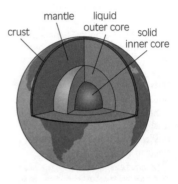

mantle
liquid outer core
crust
solid inner core

2 noun

A core is the central part of a nuclear reactor where nuclear reactions take place.

eg The reaction at the core of a reactor generates a great deal of heat.

3 noun

A core is a piece of magnetic material (e.g. soft iron) that is placed in the centre of a coil of wire through which an electric current is being passed. It increases the intensity of the magnetic field.

eg The core needs to be connected to the circuit to allow the electricity to flow.

➡ **crust, electromagnet, mantle**

correlation ???

noun

Correlation is the connection between two variables, which can be found by drawing a scatter graph.

eg This is a scatter graph showing a negative correlation.

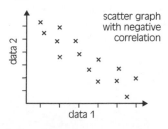

scatter graph with negative correlation

data 2

data 1

➡ **line of best fit, scatter graph**

corrosive

adjective

Something that is corrosive tends to eat away at other materials.

eg You will find the hazard label for corrosive chemicals on a bottle of sulphuric acid.

➡ **hazard label**

counterbalance

1 noun

A counterbalance is a weight or force that balances or cancels out another force.

eg In the diagram, the small weight is acting as a counterbalance to the large one.

2 verb

To counterbalance means to balance out a weight or force.

eg In this diagram, the force on the right-hand side counterbalances the force on the left-hand side.

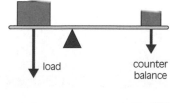

load

counter balance

crust

noun

The crust is the outer layer of the Earth.

eg The Earth's crust is between 10 km and 50 km thick.

➡ **core, mantle**

crystal

noun

A crystal is a solid which is made up from a regular arrangement of particles. The result is a solid with a regular shape. Crystals are formed by a process of crystallisation either cooling a melt or by cooling a hot, saturated solution. Small crystals are formed during rapid cooling and large crystals by slow cooling.

eg A salt crystal is cubic in shape. Salt crystals are formed when a saturated solution of salt is allowed to stand in a warm place.

cubic centimetre [cm³]

noun **? ? ?**

A cubic centimetre is a volume equal to 1 centimetre cubed.

1 cm
1 cm 1 cm

eg What is the volume of a block 4 cm × 3 cm × 2 cm?

A cubic centimetre is the same volume as 1 millilitre (1 ml).

current [I]

noun

An electric current is the flow of an electric charge through a conductor due to a potential difference.

eg The unit of current is the ampere or amp.

cuticle

noun

The waxy layer on the surface of a leaf is called the cuticle.

eg The cuticle prevents too much water evaporating from the surface of the leaf.

➡ leaf

data   **???**

plural noun

Data are information obtained especially by observation of an experiment.

eg Data were collected from the experiment and used to draw a graph.

fi The word 'data' comes from a Latin word dare, meaning 'things given'. Although you will commonly come across the word 'data' being used as a singular noun, it is more correct to use it as a plural.

data logger   **???**

noun

A data logger is a device linked to a computer to collect and process data.

eg A data logger can be used to measure the amount of light passing through a solution during an experiment.

data reliability   **???**

noun

If an experiment is repeated several times and the results are very similar, this suggests data reliability. This is an important aspect of evaluation.

eg If you compare your results with those of other students in the class and your results are similar, this suggests that the data are trustworthy and there is data reliability.

➡ **evaluation**

data search   **???**

noun

You may need to look up information when carrying out an investigation, and this is called a data search.

eg A data search may be made using books, CD-ROMs or the Internet.

datum

➡ **data**

deci-   **???**

prefix

The prefix 'deci-' means 'one-tenth'.

eg A decimetre is one-tenth of a metre. How many centimetres is that?

deficiency

noun

Deficiency means not having a sufficient amount. It is often used with reference to the diet.

eg Sailors in the 18th century often suffered from a disease called scurvy. This caused bleeding gums, muscle weakness and, ultimately, death. In 1847 Dr James Lind found out that lime juice could prevent scurvy. We now know that scurvy is due to a deficiency of vitamin C.

degree [°]

1 noun   **???**

A degree is the unit of measurement of temperature.

eg The temperature of the room was 22 degrees centigrade.

2 noun

A degree is also a unit of measurement of angles.

eg One complete revolution is 360 degrees.

density

noun

The density of a substance is the mass per unit volume. A substance with a high density has particles that are close together, and a substance with a low density has particles that are widely spaced.

eg The density of gold is 19.32 g/cm^3. Gold is the metal with the highest density. Will gases have high or low densities?

🔍 The word 'density' comes from the Latin word *densus,* meaning 'thick'.

dependent variable ???
noun

A dependent variable is a variable you do not change during an experiment. Instead, it changes as a result of other changes you make.

eg In an experiment comparing the mass of salt that dissolves in a fixed amount of water at different temperatures, you change the temperature of the water used. The mass of salt that dissolves is not controlled by you and is the dependent variable.

➡ **independent variable, variable**

deposit
noun

A deposit is a solid that has settled at the bottom of a liquid. Deposit is also the word used to refer to a layer of coal, oil or mineral naturally occurring in a rock.

eg When a bottle of limewater is left to stand, a white deposit sinks to the bottom of the bottle.

develop a technique
verb phrase ???

When you develop a technique, you learn how to carry out an action or investigation quickly, reliably and efficiently.

eg The class were asked to develop a technique for collecting woodlice safely.

di- ???
prefix

The prefix 'di-' means 'two'.

eg Carbon dioxide, CO_2, contains one carbon atom and two oxygen atoms.
➡ **bi-**

diaphragm
noun

The diaphragm is the part of the body between the abdomen and the thorax in mammals.

eg The elephant's diaphragm was enormous.
➡ **abdomen, thorax**

diffusion
noun

Diffusion is the random movement of particles from a region where they are in high concentration to a region where they are in low concentration. Diffusion takes place faster in gases than in liquids. In biological systems, diffusion is important for transferring substances over short distances. In the lungs, oxygen and carbon dioxide are transferred by diffusion.

eg The diagram shows diffusion of bromine.

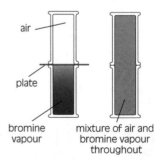

air

plate

bromine vapour

mixture of air and bromine vapour throughout

🔍 The words 'diffuse' (verb) and 'diffusion' come from the Latin word *diffundere,* meaning 'to move in all directions'.
➡ **gaseous exchange, surface area**

A B C D E F G H I J K L M N O P Q R S T U V W X Y Z

digestion

noun

The process in which nutrients in food are broken down and absorbed in the digestive system is called digestion.

eg Food cannot be used by the body directly. It must be broken down by digestion into chemicals the body can use.

➡ **digestive system**

digestive system

noun

The organs of the body (mouth, stomach, small and large intestines, etc.) that are responsible for the breaking down and absorption of food make up the digestive system.

en Chemicals such as glucose pass through the walls of the digestive system by the process of diffusion.

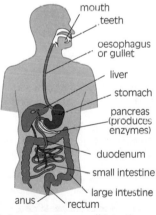

mouth
teeth
oesophagus or gullet
liver
stomach
pancreas (produces enzymes)
duodenum
small intestine
large intestine
anus
rectum

➡ **large intestine, small intestine, stomach**

dispersal (verb: disperse)

noun

Dispersal is the process whereby gametes, eggs, seeds or offspring move away from their parents into other areas. By doing this, overcrowding and competition with parents are avoided.

eg There are different methods of dispersal, e.g. by wind and water (called passive dispersal) or by their own movement (active dispersal).

🔧 The word 'dispersal' comes from the Latin word *dispergere*, meaning 'to scatter widely'.

dispersion

noun

The splitting of white light into a spectrum, for example when it is passed through a prism, is called dispersion.

eg The diagram shows how white light can be split up by dispersion.

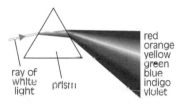

ray of white light   prism

red
orange
yellow
green
blue
indigo
violet

displacement

noun

The straight line distance between two points is called the displacement.

eg A person runs 10 m along a path. Their displacement is 10 m. They then run back. Their displacement is now zero.

displacement reaction

noun

A displacement reaction is a chemical reaction in which a metal replaces a less reactive metal in a compound.

eg Long lengths of railway track are welded together by a displacement reaction involving iron(III) oxide and aluminium. This is called the Thermit Reaction. Aluminium is more reactive than iron so it displaces the iron.

iron(III) oxide + aluminium →
aluminium oxide + iron

$$Fe_2O_3 + 2Al \rightarrow Al_2O_3 + 2Fe$$

dissipation

noun

Dissipation occurs when something is scattered or broken up and thus disappears.

eg The dissipation of the smoke did not take very long.

dissolve

verb

To dissolve is to go into a solution and disappear, or to become a liquid.

eg When salt is added to water, the salt will dissolve and form a salt solution. The salt can be recovered by evaporation.

➡ **solubility, solute, solution, solvent**

distillation

noun

Distillation is a method used to separate a volatile substance with a low boiling point from involatile impurities. The process involves boiling followed by condensation.

eg Pure water can be separated from salt water by distillation. The solution is boiled and the steam condensed to produce pure water. The diagram shows apparatus suitable for distillation.

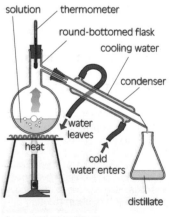

➡ **fractional distillation**

distribution

noun

A distribution is a collection of measurements or data.

eg The chart shows the distribution of heights of men in a sample.

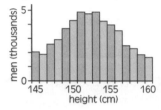

dormant

adjective

An organism is said to be dormant when it is in a resting state.

eg Seeds may remain dormant in soil for years before they start to germinate.

ℹ The word 'dormant' comes from the Latin word *dormire*, which means 'to sleep'.

➡ **hibernation**

drag

noun

When an object is moving through a fluid, liquid or gas, there is a resistive force against the movement of the object. This is called the drag.

eg When a ball bearing falls through motor oil, it soon reaches its maximum velocity, called the terminal velocity. This is the point at which the drag exactly balances the weight of the ball bearing.

Ball bearing dropping in oil

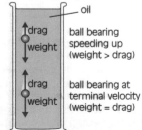

drug

noun

A chemical that alters the way the body works is called a drug. There are different types of drug including:

- stimulants that speed up the brain and make you more alert, e.g. caffeine, Ecstasy and amphetamines
- sedatives that slow down the brain and make you sleepy, e.g. barbiturates
- painkillers that affect parts of the brain to remove the feeling of pain, e.g. aspirin and paracetamol.

Alcohol is a widely used drug. Cigarettes are another. What makes them addictive?

ductile (*duck-tile*)

adjective

Metal is said to be ductile if it can be drawn out into a thin wire.

Copper is ductile and can be made into thin wire.

dynamics

noun

Dynamics is the branch of mechanics that deals with motion and the forces that produce motion.

Design engineers producing a new car need to compare the dynamics of different designs.

The word 'dynamics' comes from the Greek word *dynamis*, which means 'power'.

dynamo

noun

A dynamo is a device for transforming mechanical energy into electrical energy.

A dynamo is often found on a bicycle where it provides the electrical energy for the lamps.

➡ generator

A
B
C
D
E
F
G
H
I
J
K
L
M
N
O
P
Q
R
S
T
U
V
W
X
Y
Z

ear

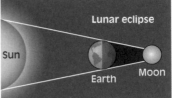

noun

The ear is the hearing organ. It collects sound vibrations and turns them into electrical signals that are sent to the brain.

eg The diagram below shows the cross-section of the ear.

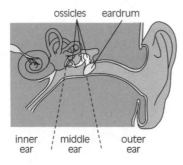

ossicles   eardrum

inner / middle \ outer
ear / ear \ ear

eclipse

noun

An eclipse occurs when one astronomical body is partially or totally concealed or obscured by another.

eg A solar eclipse takes place when the Moon passes in front of the Sun, as seen from the Earth. A lunar eclipse occurs when the Moon passes into the shadow of the Earth, becoming dim until it emerges from the shadow.

The diagrams show solar and lunar eclipses (not to scale).

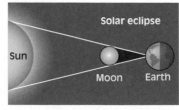

Solar eclipse

Sun
Moon   Earth

Lunar eclipse

Sun
Earth   Moon

i The word 'eclipse' comes from the Greek word *ekleipsis*, which means 'failure to appear'.

ecosystem

noun

An ecosystem is a community of living things and their relationships with their surroundings.

eg An ecosystem may be aquatic (e.g. lake or river) or terrestrial (e.g. forest or grassland).

egg

noun

Egg is another name for ovum.

eg The egg is the unfertilised female gamete.

➡ **ovum**

electric generator

noun

An electric generator is a machine that produces electricity.

eg The electric generator provided all the electricity in the house.

➡ **generator**

electromagnet

noun

An electromagnet is made by passing an electric current through a wire that is coiled around an iron bar.

eg At least one electromagnet is used for switches and electric bells.

element

noun

An element is a pure substance that cannot be split up into simpler substances. There are 109 known elements, of which 95 occur naturally.

eg All elements can be represented by a symbol. The symbol for an element is one or two letters. For example, the symbol for the element magnesium is Mg.

elodea

noun

Elodea is a type of waterweed used in photosynthesis experiments.

eg When light shines on elodea, oxygen gas is produced.

- bubble of liquid
- scale
- oxygen gas
- light
- elodea

embryo

noun

An embryo is a stage in the development of a plant or animal.

eg In humans, the term 'embryo' refers to the first seven weeks after fertilisation. From the eighth week onwards it is called the fetus.

emphysema (em-fa-see-ma)

noun

Emphysema is a disease caused by the breakdown of the walls of the alveoli, thus reducing gaseous exchange.

eg Emphysema reduces the surface area of the alveoli.

bronchiole

alveolus

In a healthy person **In a person with emphysema**

The word 'emphysema' comes from the Greek word *emphusēma*, which means 'to swell'.

energy

noun

Energy is the ability to do work and is measured in units called joules.

eg Energy can be potential energy (due to vertical position) or kinetic energy (due to movement). A stone dropped over a cliff has potential energy due to its position. When the stone reaches the bottom of the cliff the stone has lost its potential energy but has gained kinetic energy.

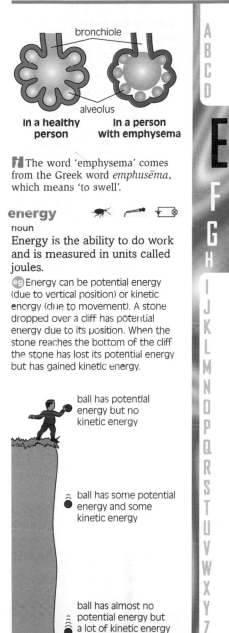

ball has potential energy but no kinetic energy

ball has some potential energy and some kinetic energy

ball has almost no potential energy but a lot of kinetic energy

energy transfer

noun

Energy cannot be created or destroyed. It is possible, however, for energy transfer to occur from one form to another.

eg In an electric kettle, electrical energy is transferred to heat energy. The diagrams show two devices in which energy transfer takes place. What energy transfer occurs in each device shown?

solar-powered calculator

loudspeaker

environmental conditions

noun

The surroundings in which plants, animals and insects live are called their environmental conditions.

eg The environmental conditions of the moths were dark and warm.

enzyme

noun

An enzyme is a chemical found in the digestive system that is used to break down large particles into smaller pieces. Enzymes act as biological catalysts. They are proteins.

eg Starch is broken down by the enzyme amylase.

🔏 The word 'enzyme' comes from the Greek word *zyme*, meaning 'leaven', which is a substance added to dough to make it rise.

epidemic

noun

An epidemic is a sudden outbreak of an infectious disease which spreads rapidly and affects a large number of people, plants or animals in a particular area for a limited period of time.

eg Following an epidemic of cholera in London, sewers were planned.

🔏 The word 'epidemic' comes from the Greek words *epi* and *demos*, meaning 'among the people'.

equation

noun

An equation is a summary of a chemical reaction, written either in words or in symbols.

eg The reaction of magnesium and oxygen can be summarised by the following word and symbol equations.

magnesium + oxygen $\rightarrow$ magnesium oxide

$$2Mg + O_2 \rightarrow 2MgO$$

➡ symbol equation, word equation

erosion

noun

Erosion is the breaking down and transport of rocks involving movement of the sea, rivers, glaciers and the wind, etc.

eg Running water causes the erosion of rocks.

➡ rock cycle, weathering

erupt

verb

A volcano is said to erupt when lava, ash and gases are pushed out.

eg Gases that escape when volcanos erupt can alter the composition of the atmosphere.

🔏 The word 'erupt' comes from the Latin word *erumpere*, which means 'to break out'.

etiolation (*e-tee*-o-lay-shun)

noun

Etiolation takes place when a plant is grown in the absence of light.

eg Etiolated plants, i.e. plants that have suffered from etiolation, are yellowed and spindly. Light is needed to produce chlorophyll.

The word 'etiolation' comes from the French word *étoiler*, which means 'to become pale'.

evaluation (verb: evaluate)

noun ? ? ?

Evaluation occurs when a judgement is made about the worth of something.

eg When you are doing an evaluation of an experiment you have done, you should:

● Look at the quality of the data. Are there any anomalous results? Are there sufficient results? Are the results reliable?

● Try to suggest improvements to the procedure you have followed.

➡ **anomalous result, data reliability**

evaporation

(verb: evaporate)

noun

Evaporation is the process whereby a liquid is turned into a gas below its boiling point.

eg When an uncovered dish containing water is left on a windowsill, evaporation of the water takes place.

➡ **boiling temperature, change of state**

evidence ? ? ?

noun

Evidence is the information that gives sufficient grounds for believing something.

eg Evidence from the experiment confirms the prediction I made.

excess ? ? ?

adjective

An excess amount is an amount greater than that which is actually needed.

eg When a lot of salt is added to water, the excess salt that cannot dissolve sinks to the bottom.

exercise

noun

Exercise is physical training or exertion with a view to improving health.

eg Taking exercise increases the pulse rate and breathing rate.

exhale

verb

The word 'exhale' means to breathe out.

eg The opposite of exhale is inhale. The two processes together move air in and out of the lungs.

➡ **breathing, inhale**

expansion

noun

Expansion is the increase in size of a constant mass of a body caused by increasing its temperature or its internal pressure.

eg When a piece of metal is heated, it expands. This expansion could cause problems with long pieces of railway track. Special measures need to be taken to stop railway tracks buckling.

expire

verb

To expire means to breathe out.

eg Expire is the opposite of inspire.

➡ **breathing, exhale**

A B C D E F G H I J K L M N O P Q R S T U V W X Y Z

explain ???

verb

To explain something is to make it easy to understand.

🔵 Explain the stages involved in the rock cycle.

🔶 The word 'explain' comes from the Latin word *explanare*, which means 'to make flat'.

explosive

adjective

Something is described as explosive if it is likely to burst due to a sudden and violent increase in pressure that also gives out heat and light.

🔵 Dynamite is an explosive substance first made by Alfred Nobel. What was named after Nobel?

➡ **hazard label**

extrapolate ???

verb

To extrapolate is to predict results beyond the extent of the given values.

🔵 Steven finds out that a force of 10 N extends a spring by 1 cm and that a force of 20 N extends the same spring by 2 cm. He can extrapolate this data and propose that a 30 N force would extend the spring by 3 cm.

fair test ? ? ?
noun

In a fair test, you should attempt to keep all the variables constant except the one you are investigating.

eg To ensure you have a fair test of an investigation into the time taken for different masses of sugar to dissolve in water, you should vary the mass of sugar used. You should keep the following constant:

- same volume of water
- water at the same temperature
- same type of sugar
- same amount of stirring.

fat
noun

Fat is a substance found in food that acts as an energy store. Excess carbohydrates and proteins are converted to fat for storage.

eg Food can be tested for fat by dissolving the food in ethanol and then pouring the solution into water. If the solution goes cloudy-white, a fat is present.

feature
noun

The word 'feature' means characteristic.

eg An important feature of a sperm cell is a long tail which makes it very mobile.

H The word 'feature' comes from a 14th-century French word, *faiture*, meaning 'bodily form'.

fern
noun

A fern is a flowerless, feathery-leaved plant that reproduces by spores.

eg Where are you likely to find a fern growing?

fertilisation (verb: fertilise)
noun

Fertilisation is the fusing of the nuclei of two gametes (ovum and sperm) to form a zygote.

eg The diagram below summarises the process of human fertilisation.

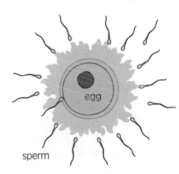

egg

sperm

➡ **gamete, ovum, sperm, zygote**

fertiliser
noun

A fertiliser is a natural or synthetic substance containing compounds of nitrogen, potassium and phosphorus, which can compensate for deficiencies in soil.

eg If a fertiliser is over-used, it can cause water pollution in lakes and rivers.

fetus/foetus (*fee-tus*)

noun

The fetus is a stage in the development of the embryo in a mammal.

eg The human embryo is usually called a fetus after eight weeks of development. At this stage, the limbs and external features such as the head can be seen in a scan.

The diagram shows a fetus developing.

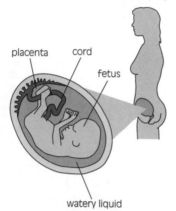

placenta   cord

fetus

watery liquid

fibre (*fye-ber*)

noun

Fibre, sometimes called roughage, is an important part of the diet. It is not digested in the digestive system but helps in the production of faeces and prevents constipation. It also reduces the risk of bowel cancer.

eg Fruits, vegetables and cereals provide good sources of fibre.

filter

1 noun

A material, such as paper or sand, that allows liquids to pass through but catches solid particles is called a filter.

eg The coffee grounds were left in the filter paper but the liquid coffee passed through.

2 verb

To filter means to remove or separate solids from a liquid.

eg The scientist had to filter the mixture to catch the solid particles.

filtration

noun

Filtration is a process that can be used to separate a solid from a liquid.

eg Filtration can be used to separate a mixture of sand and salt. If they are mixed with water and then filtered, the sand remains on the filter paper but the salt solution passes through the filter paper into the basin.

beaker

sand

funnel

filter paper

basin

salt solution

flammable

adjective

A substance is said to be flammable if it easily catches alight.

eg Petrol is highly flammable.

The word 'flammable' comes from the Latin word *flammare*, which means 'to blaze'. The words 'flammable' and 'inflammable' mean the same thing – they are not opposites. The opposite of flammable and inflammable is non-flammable.

➡ hazard label

flask

noun

A flask is a container, usually made from glass, that is used in scientific experiments.

eg This diagram shows a range of different flasks.

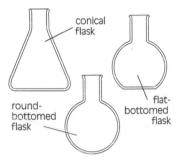

conical flask

round-bottomed flask

flat-bottomed flask

flexible

adjective

If something is flexible, it is supple and easy to bend. The opposite of flexible is rigid.

eg A piece of card is flexible.
➡ rigid

float

verb

To float means to rest on or in a liquid without sinking. An object will float in a liquid when the weight and the upthrust are equal.

eg The diagram shows the forces acting on an object that floats.

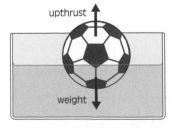

upthrust

weight

food chain

noun

A food chain is a way of showing the feeding relationships between plants and animals.

Food chains are written down using arrows. The arrow means 'eaten by' and always points to the right.

eg This is one food chain, from the producer (grass) to the end consumer (hawk):

grass → grasshopper → frog → hawk
➡ consumer, food web, producer

food poisoning

noun

Food poisoning is an illness caused by eating food or drinking water containing toxins or micro-organisms, such as salmonella.

eg Food poisoning often occurs because of errors in food preparation and storage.

food web

noun

A food web shows the feeding relationships between a number of different organisms. It consists of a number of food chains.

eg The diagram shows a food web for organisms in an oak tree.

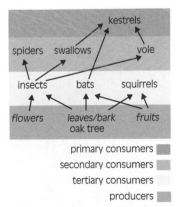

kestrels

spiders   swallows   vole

insects   bats   squirrels

flowers   leaves/bark   fruits
oak tree

primary consumers
secondary consumers
tertiary consumers
producers

A B C D E F G H I J K L M N O P Q R S T U V W X Y Z

force

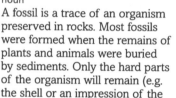

noun

A force (F) is an external agent that tends to change the state of rest or the uniform motion in a straight line of a body. The action of an unbalanced force produces acceleration of a body in the direction of the force. A force may also change the shape of an object, e.g. a spring. Force is a vector quantity having a magnitude and a direction. The unit of force is the Newton.

eg A mass of 1 kg is pulled towards the Earth by a downward force of 10 N.

🛈 The unit of force is named after Sir Isaac Newton. He was an important scientist in the 17th century.
➡ **acceleration, gravity**

force meter

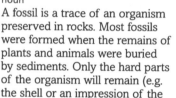

noun

A force meter is an instrument used to measure the size of a force. There is a spring inside the force meter. The greater the force, the greater the extension of the spring. Force meters are also called Newton meters.

eg This diagram shows a force meter.

formula (plural: formulae/ formulas)

noun

A formula is a general rule or relationship stated in the form of symbols.

A chemical formula shows molecules, etc., expressed in the symbols of their constituent atomic elements.

eg The chemical formula for water is H_2O.

fossil

noun

A fossil is a trace of an organism preserved in rocks. Most fossils were formed when the remains of plants and animals were buried by sediments. Only the hard parts of the organism will remain (e.g. the shell or an impression of the shell in the rock).

eg The presence of a fossil in a rock can help to date the rock. If the same fossils were found in another rock of a known age, the two rocks would both be of this age.

fossil fuel

noun

A fuel produced by the action of high temperatures and high pressures on plant and animal material over millions of years is called a fossil fuel. Examples of fossil fuels are coal, crude oil and natural gas.

eg Fossil fuels are non-renewable fuels that will eventually run out.

fractional distillation

noun

Fractional distillation is a process used to separate mixtures of liquids with different boiling points.

eg Fractional distillation produces a range of different liquids with different boiling points and uses.

The diagram shows the apparatus that can be used for fractional distillation in the laboratory.

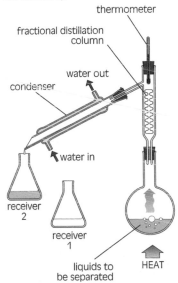

thermometer

fractional distillation column

water out

condenser

water in

receiver 2

receiver 1

liquids to be separated

HEAT

➡ change of state, distillation

freeze

verb

To freeze something is to change a liquid into a solid by reducing its temperature.

eg When water is cooled to 0°C, it will freeze and turn to ice.

The word 'freeze' comes from an Anglo-Saxon word, *freosan*.
➡ change of state

frequency

noun

The number of vibrations in a second is called the frequency.

The unit of frequency is the hertz. One hertz is equivalent to one cycle per second.

eg A human being hears sounds with a frequency between 20–15 000 Hz.

friction

noun

Friction is the force that opposes the relative motion of two bodies in contact.

eg Friction is reduced by using a lubricant such as oil. Friction is important for brake linings, soles of shoes and tyres.

The word 'friction' comes from the 16th-century Latin word *fricare*, which means 'to rub'.
➡ lubricant

fuel

noun

A fuel is any substance that releases energy when it is burned. Nuclear fuels are fuels that produce energy in a nuclear reactor when burning.

eg Some fuels are called fossil fuels, for example coal, oil and natural gas. Wood and straw are each an example of a renewable fuel.

The word 'fuel' comes from the Latin word *focus*, which means 'hearth'.

function

noun

A function is a job.

eg The function of the heart is to pump blood around the body.

fungicide

noun

A chemical that kills or limits the growth of a fungus is called a fungicide.

eg In a damp room, the walls are treated with a fungicide to stop the growth of black mould.

fungus (plural: fungi)

noun

A fungus is an organism that resembles a plant but has no leaves or roots. It contains no chlorophyll and reproduces by spores. Since it cannot produce food, it gets its food either by living off other plants or animals (parasitic) or living off dead matter (saprophytic).

eg Moulds, yeasts, mildews and mushrooms are all types of fungus.

funnel

noun

A funnel can be used for transferring a liquid from one container to another and for filtering a solid from a liquid.

eg The diagram below shows a funnel.

Gg *Gg* Gg Gg **Gg**

A
B
C
D
E
F
G
H
I
J
K
L
M
N
O
P
Q
R
S
T
U
V
W
X
Y
Z

gamete
noun

A gamete is a specialised sex cell, e.g. a sperm or ovum, which fuses with another gamete of the opposite type during fertilisation.

A human body cell contains 46 chromosomes. How many chromosomes are there in a gamete?

The word 'gamete' comes from the Greek word *gameein*, which means 'to marry'.

➡ **ovum, sperm, zygote**

gas
noun

One of the three states of matter, a gas is an air-like substance that moves freely to fill the space available.
A gas:
● has a low density
● has widely spaced particles
● is easily compressed.

Below is a diagram showing the arrangement of particles in a gas.

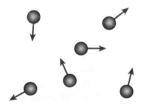

A cube of air $1\,cm \times 1\,cm \times 1\,cm$ contains 20 million million particles moving at an average speed of 500 m/s.

➡ **change of state, liquid, solid**

gaseous exchange
noun

When oxygen passes into the blood and carbon dioxide is moved out of the blood, gaseous exchange occurs.

Gaseous exchange takes place in the alveoli in the lungs by a process of diffusion.

➡ **alveolus, diffusion, lung, respiratory system**

gas pressure
noun

The particles in a gas are moving all the time in all directions. The particles collide with each other and with the walls of the container. The greater the number of collisions with the walls, the higher the gas pressure.

Raising the temperature of a gas increases the average speed of the particles. What does this do to the gas pressure?

gene *(jean)*
noun

A gene is a section of a chromosome which, either on its own or with other genes, is responsible for a particular characteristic.

The diagram shows different genes in a chromosome.

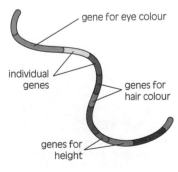

gene for eye colour

individual genes

genes for hair colour

genes for height

fi There is a chemical called phenylthiocarbamide (PTC) which about 70% of the population can taste but 30% cannot. This depends on whether one particular gene is present.

➡ chromosome

generalise
(noun: generalisation) **? ? ?**
verb

To generalise is to form a general statement or rule from the evidence available.

eg Having tested some metals with dilute acids, we can generalise that when a metal reacts with dilute acid, hydrogen gas is produced.

generator
noun

A generator turns mechanical energy into electrical energy. It works on the principle that a magnetic field rotating inside a coil generates a voltage in that coil. It is commonly used in a bicycle dynamo or a power station.

eg A bicycle dynamo is a small generator.

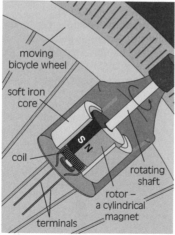

moving bicycle wheel

soft iron core

coil

N S

rotating shaft

rotor – a cylindrical magnet

terminals

Why does the bicycle lamp shine brighter when you pedal faster?

genetic modification
noun

Genetic modification occurs when genetic material is deliberately changed using biochemical techniques. Genetic modification can be used to produce new varieties of plants, animals or bacteria.

eg Genetic modification has enabled scientists to produce human insulin, human growth hormone and vaccines.

germ
noun

The word 'germ' is commonly used to describe microbes that cause disease. About 200 years ago, Louis Pasteur developed the theory that diseases are caused by germs.

eg Pasteur showed that soup decayed when in contact with air but not when out of contact with air. He concluded that there were germs in the air that caused this change.

➡ bacteria, micro-organism, virus

germination
(verb: germinate)
noun

Germination is the beginning of the growth of a seed. The process begins with the seed taking up water. Then the root usually emerges from the seed, followed by the shoot. Germination is finished when the first true leaves are formed.

eg The conditions needed for germination are water, oxygen and a suitable temperature (usually above 5°C and below 45°C).

fi The verb 'to germinate' and the noun 'germination' both come from the Latin word *germinare*, which means 'to sprout'.

gestation

noun

The time between implanting the embryo in the uterus of a mammal and birth is called gestation.

eg In the case of a human, the gestation time is nine months, but in an elephant it is 18 to 22 months.

giga-

prefix

'Giga-' is a prefix meaning 1 000 000 000.

eg A gigawatt is 1 000 000 000 watts.

global warming

noun

Global warming is a rise in the average temperature of the Earth's surface. It is thought to be caused by the greenhouse effect. Gases in the atmosphere such as carbon dioxide and methane prevent infrared radiation from the Sun escaping from the Earth.

eg What processes increase the level of carbon dioxide in the atmosphere and increase global warming?

glucose (*gloo*-kose)

noun

Glucose is a substance produced by photosynthesis in a plant and is used as an energy source in respiration. It is a carbohydrate with a chemical formula $C_6H_{12}O_6$.

eg Like all carbohydrates, glucose is made up from three elements. One is carbon. What are the other two?

➡ **carbohydrate, photosynthesis**

graft

verb

To graft means to remove a piece of living tissue from a donor or from a patient's own body and transplant it to another part of the body.

eg One example of grafting is found in gardening. When propagating a rose tree, a bud or shoot (called a scion) from one plant is inserted (grafted) into another (called the stock) and the two grow together. The resulting rose tree has some advantages of both the original plants.

gram [g]

noun

A gram is a unit of mass in the metric system.

eg Ella weighed a crucible and found it has a mass of 34.82 grams.

granite

noun

Granite is an igneous rock formed when magma crystallises slowly inside the Earth. It usually consists of large crystals and is frequently made of the minerals quartz, feldspar and mica.

eg Granite is an extremely hard-wearing rock commonly used for kitchen worktops.

graph

???

noun

A diagram showing the relationship between two variable quantities is called a graph. Each quantity is measured along an axis.

eg The graph below shows time along the x-axis (horizontal axis) and distance along the y-axis (vertical axis).

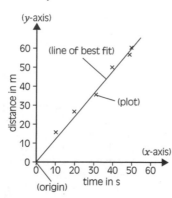

(y-axis)

60 — (line of best fit)

50

40 distance in m

30 — (plot)

20

10

(x-axis)

0 10 20 30 40 50 60

time in s

(origin)

➡ axis, dependent variable, independent variable

gravitational attraction

➡ gravitational force

gravitational force

noun

A gravitational force is a force of attraction between objects.

Although they act between all objects, these forces are only noticeable if one of the objects is massive.

eg The Sun exerts a gravitational force on the planets, which keeps them in their orbits around the Sun.

➡ attraction, weight

gravity

noun

The force of gravity on the Earth is the force of attraction between any object in the Earth's gravitational field and the Earth itself.

eg When an object is dropped, the force of gravity causes it to fall back to Earth.

➡ force

growth

noun

Growth means, specifically, an increase in the size and dry weight (i.e. weight without water) of an organism. It is linked with cell division. If a cell is put into water and it expands, it is still just one cell – it is not growth.

eg A seedling was 1.2 cm high last week and is 1.6 cm high today. What was its growth in one week?

habitat
noun
A habitat is the natural home of a plant or animal from where it can obtain all (or nearly all) of its needs. The number of different types of habitat within the Earth's ecosystem is enormous.

eg The habitat of penguins is the icy wastes of Antarctica.

The word 'habitat' comes from the Latin word *habitare*, which means 'to dwell'.
➡ ecosystem

haemoglobin
(*hee-ma-glow-bin*)
noun
Haemoglobin is the protein in red blood cells that carries oxygen to cells around the body. There is an iron atom at the centre of each haemoglobin molecule.

eg Oxygen combines with haemoglobin to form a compound – oxyhaemoglobin – that easily breaks down to reform haemoglobin and release oxygen.

The prefix 'haemo-' comes from the Greek work *halma*, meaning 'blood'.

harmful
adjective
Something is said to be harmful if it can cause physical or mental injury.

eg Dilute solutions of some chemicals can be harmful.
➡ hazard label

hazard
noun
Something that might cause injury or loss is called a hazard.

eg The water was a hazard to young children.

hazard label
noun
A hazard label is one of a series of warning labels used to indicate potential dangers in handling substances. These symbols are used throughout the world.

eg You will see a hazard label on many products in the kitchen.

The most common hazard labels are shown below.

highly flammable | harmful
toxic | corrosive

health
noun
Health is a state of physical, mental and social wellbeing.

eg Good health also assumes freedom from illness and pain.

heart
noun
The heart is the muscular organ in vertebrates that pumps blood around the body.

A
B
C
D
E
F
G

H

I
J
K

L
M
N
O
P
Q
R
S
T
U
V
W
X
Y
Z

eg Arteries carry oxygen-rich blood from the heart around the body, and veins carry oxygen-poor blood back to the heart.

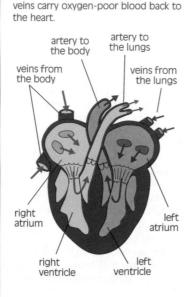

artery to the body
artery to the lungs
veins from the body
veins from the lungs
right atrium
left atrium
right ventricle
left ventricle

heartbeat

noun

The alternate contraction and relaxation of the heart muscle produces a beating sound called the heartbeat.

eg Doctors use a stethoscope to listen to the heartbeat of patients.

heat

noun

Heat is the energy transferred as a result of a difference in temperature.

eg It is important not to confuse heat with temperature. There is a lot of heat stored in a bath of cold water but the water is at a low temperature.

➡ **temperature**

heliocentric

adjective

Something is said to be heliocentric if the Sun is at its centre.

eg Planets such as Venus and Mercury have a heliocentric orbit around the Sun.

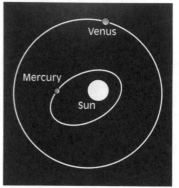

Venus
Mercury
Sun

🔎 The prefix 'helio' comes from the Greek word *helios*, which means 'the Sun'.

herbicide

noun

A herbicide is a chemical that kills plants.

eg The gardener used the herbicide to kill the dandelions in her lawn. This needs to be a selective weedkiller that kills dandelions without killing the grass.

➡ **weedkiller**

herbivore

noun

An animal that feeds on grass and other plants is called a herbivore.

eg A cow is a herbivore.

hereditary

adjective

Something is said to be hereditary if it is passed on genetically from one generation to the next. Some diseases are hereditary.

Sickle-cell anaemia is a hereditary disease.

The word 'hereditary' comes from the Latin word *hereditas*, which means 'inheritance'.

hibernation
(verb: hibernate)

noun

Some animals pass the winter in a dormant state called hibernation. When they do this, all the body processes slow down. There is a fall in body temperature, breathing and heart rate.

Tortoises go into hibernation over the winter.

The word 'hibernate' comes from the Latin word *hibernus*, meaning 'wintry'.

hour ???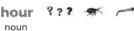

noun

An hour is a measurement of time consisting of 60 minutes.

There are 24 hours in a day.

humidity

noun

The humidity is the percentage of water vapour in the atmosphere.

In countries like Thailand, the humidity is very high.

hydraulic (*high-**draw**-lick*)

adjective

Something is said to be hydraulic if it is operated by pressure transmitted through a pipe by a liquid. Brake systems are often hydraulic.

The diagram shows a simplified representation of the hydraulic braking system in a car. Why is the system less effective if air is trapped in the pipe?

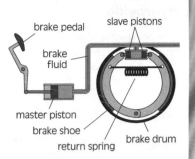

The word 'hydraulic' comes from two Greek words, *hydor* and *aulos*, meaning 'water' and 'pipe'.

hydrochloric acid [HCl]
(*high-drow-**claw**-ric*)

noun

Hydrochloric acid (formula HCl) is a mineral acid that can be made from salt.

Hydrochloric acid can be used to make salts called chlorides.

➡ chloride

hydrogen [H] (*high-drow-gen*)

noun

Hydrogen is a flammable, colourless gas that is the lightest and most common element in the universe. When a lighted splint is put into a test tube filled with hydrogen gas, there is a squeaky pop.

Hydrogen is formed when a metal reacts with water or an acid.

Hydrogen was first discovered by Henry Cavendish in 1767. Its name comes from two Greek words, *gennaein* and *hydro-*, meaning 'made from water'. Hydrogen was first produced from water.

A B C D E F G **H** I J K L M N O P Q R S T U V W X Y Z

hydroxide

noun

A hydroxide is a compound of a metal with –OH groups.

eg Sodium hydroxide, potassium hydroxide and calcium hydroxide are three common hydroxides. They are strong alkalis.

➡ **alkali, caustic**

hygiene (*hi-jean*)

noun

Hygiene is the science of preserving health and the prevention of the spread of disease.

eg It is important to keep good hygiene in the kitchen.

🔲 The word 'hygiene' comes from the Greek word *hugieine*, which means 'health'.

hypothesis (plural: hypotheses) (*hie-poth-i-siss*)

noun

A hypothesis is a statement or assumption made without experimental evidence that can be proved or disproved by reference to evidence or facts.

eg The ancient Greek, Democritus, produced a hypothesis that all matter was made up of atoms.

➡ **scientific method, theory**

-ide

suffix

A compound made up from two elements has a name with the suffix '-ide'.

eg Sodium chloride is a compound of sodium and chlorine only.

idea

? ? ?

noun

An idea is a thought formed in the mind.

eg An idea about the distribution of plants in a field was the basis of an investigation.

identify

? ? ?

verb

To identify means to recognise someone or something.

eg You could use a key to identify small organisms in leaf litter.

igneous (*ig-nee-us*)

adjective

A rock formed by the crystallisation of magma is said to be igneous.

eg The texture of igneous rocks is determined by how fast the magma cools.

The word 'igneous' comes from the Latin word *ignis*, which means 'fire'.
➡ **basalt, granite, magma, metamorphic, rock cycle, sedimentary**

image

noun

An image is a picture or an appearance of a real object produced by light rays that pass through a lens or which are reflected by a mirror.

eg The image produced by a camera is upside down (inverted) and is an example of a real image. A real image is one that can be caught on a screen.

An image in a mirror is a virtual image as it is formed behind the mirror. Mirror images are upright.

The word 'image' comes from the Latin word *imago*, which means 'likeness'.

immunisation

noun

Immunisation is the artificial immunity to infection provided by injecting a treated antigen.

eg Edward Jenner showed in 1770 that immunisation of a boy with cowpox prevented the boy contracting smallpox.

immunity

noun

Immunity is the protection that organisms have against invading micro-organisms, such as bacteria and viruses, and against cancerous cells. This protection is provided by white blood cells. The skin also provides a barrier against infection.

eg People who have had chickenpox as a child have some immunity against further infection.
➡ **antigen**

independent variable

noun
? ? ?

An independent variable is a variable that you change during an experiment. When it changes, other variables, called dependent variables, change.

eg In an experiment comparing the mass of salt that dissolves in a fixed amount of water at different temperatures, you change the temperature of the water used. This is the independent variable. The mass of salt that dissolves is not controlled by you and is the dependent variable.
➡ **dependent variable, variable**

indicator
noun

An indicator is a dye or mixture of dyes that changes colour depending upon the pH of a solution.

eg Litmus is an indicator produced from a lichen found in the Arctic Circle. It is used as solution or soaked into a piece of paper. It changes to red in acid and blue in alkali.

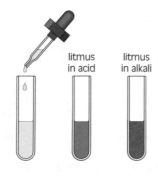

litmus in acid   litmus in alkali

➡ universal indicator

infection
noun

When the body is attacked by micro-organisms, it is said to have an infection.

eg The infection made the doctor feel quite ill.

infectious disease
noun

An infectious disease is a disease caused by bacteria, viruses or other micro-organisms. It is usually transmitted through the air or by water, etc.

eg Influenza is an infectious disease.

❗ There is often confusion between the words 'infectious' and 'contagious'. An infectious disease is transmitted through the air, for example, and a contagious disease is spread by touching.

infrared [ir]
adjective

Radiation is said to be infrared when it is between the red end of the visible spectrum and microwaves and radiowaves.

eg An infrared camera can be used to produce images at night.

inhale
verb

To inhale is to breathe in.

eg The opposite of inhale is exhale. The two processes together move air in and out of the lungs.
➡ breathing, exhale, inspire

inherit
verb

A person receiving genetically transmitted characteristics from a parent is said to inherit these characteristics.

eg A new baby will inherit all his or her features from his or her mother and father.

inoculation
noun

An inoculation is the introduction of micro-organisms into a body in order to give that body immunity to disease.

eg As a result of inoculations, smallpox has been wiped out from the world.

insecticide
noun

A chemical that is used to kill insects is called an insecticide.

eg DDT was used as an insecticide until it was found to be affecting some organisms, e.g. birds of prey. It is no longer used.

❗ DDT is the abbreviation for dichlorodiphenyltrichloroethane.

insoluble
adjective

A substance that does not

dissolve in a particular solvent is said to be insoluble.

🐞 Sand is insoluble in water.

inspire
verb
To inspire is to draw breath into the lungs, or to inhale.

🐞 Inspire is the opposite of expire.

ℹ️ The word 'inspire' comes from the Latin word *inspirare*, which means 'to breathe into'.

➡️ **breathing, inhale**

insulator
noun
A material that reduces the amount of heat, electricity or sound energy moving through it is called an insulator.

🐞 Electric cables are coated with a plastic material called PVC, which is a good electrical insulator. The cavity between the inner and outer walls of a house may be filled with a solid foam. This acts as a good insulator for heat.

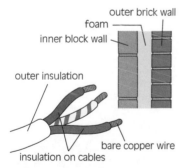

outer brick wall
foam
inner block wall

outer insulation

bare copper wire
insulation on cables

ℹ️ The word 'insulator' comes from the Latin word *insula*, which means 'island'.

interdependence
noun
Interdependence is the way in which living organisms depend on each other in order to remain alive, grow and reproduce.

🐞 Bees feed on the nectar from flowers, and at the same time they pollinate the flowers. The bees and flowers have an interdependence.

interpret ???
verb
To interpret is to explain the meaning of something and to come to a conclusion.

🐞 The scientist could interpret his results more easily from a graph.

intestine
noun
Tubes between the stomach and the anus are called the intestine.

🐞 In the intestine the digestion process is completed and food chemicals are absorbed into the bloodstream.

ℹ️ The word 'intestine' comes from the Latin word *intestinus*, which means 'internal'.

➡️ **large intestine, small intestine**

invertebrate 🐞
noun
An animal without a backbone is called an invertebrate. Of the approximately 1 000 000 different species of animals, about 95% are invertebrates.

🐞 Which of the animals below are invertebrate and which are vertebrate?

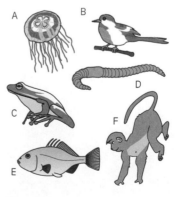

A B
C D
E F

➡️ **vertebrate**

A B C D E F G H I J K L M N O P Q R S T U V W X Y Z

irreversible

adjective

An irreversible reaction is a reaction after which the products cannot be turned back to the reactants. Most reactions are irreversible.

eg When a piece of wood is burned, wood ash, carbon dioxide and water vapour are formed. This reaction is irreversible because it is impossible to get the wood back again.

-ite

suffix

A salt with a name ending in '-ite' contains oxygen, but it contains less oxygen than a salt with a name ending in '-ate'.

eg Sodium nitrate, $NaNO_3$, and sodium nitrite, $NaNO_2$, are two salts. They both contain oxygen but sodium nitrite contains less oxygen than sodium nitrate.

Jj

joint

noun

A joint is a point where two bones meet in a vertebrate. In some cases, for example in the skull, there is no movement between the two bones. In most cases, however, there is movement.

Two types of joint are the hinge joint (e.g. elbow or knee) and the ball and socket joint (hip and shoulder).

Hinge joint (knee)

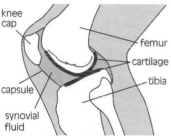

Ball and socket joint (hip)

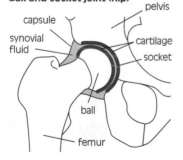

joule (*jool*) [J]

noun

The joule is the SI unit of work and energy.

Tim weighs 600 N. He climbs some stairs, the vertical height of which is 1.5 m.

The work he does = 600 × 1.5
= 900 Joules.

The unit of energy and work is named after James Joule, a famous 19th-century scientist.

Kk

kelvin [K]
noun

The kelvin is an SI unit for temperature.

eg A Celsius temperature can be turned into a temperature on the Kelvin scale by adding 273. What is the boiling point of water on the Kelvin scale?

i The unit is named after the scientist, Lord Kelvin.

key
noun

A key is used to identify an organism. An investigator is given a set of statements. They choose the best statement and then move on to the next set of statements. At each stage they eliminate some organisms and so hopefully finish up naming the organism correctly.

eg The picture shows four arthropods. Follow the key to identify them.

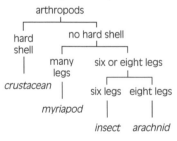

kidney
noun

The kidney is a bean-shaped organ found in the abdomen cavity. In vertebrates, the two kidneys are responsible for water regulation, excretion of waste products from the body and maintenance of the composition of the blood.

eg Waste products from blood are removed in each kidney and excreted in urine.

kilo-
prefix

The prefix 'kilo-' means 'one thousand times'.

eg A kilometre is 1000 metres.

kilogram [kg]
noun

A kilogram is a mass of 1000 grams.

eg Fruit and vegetables are sold in kilograms. One kilogram is about 2.2 pounds.

kilometre [km]
noun

A kilometre is a measurement of length equal to 1000 metres.

eg A kilometre is approximately five-eighths of a mile.

kinetic energy
noun

Kinetic energy is the energy possessed by a moving object. It can be calculated using the formula $KE = \frac{1}{2} mv^2$.

eg Particles in all substances possess kinetic energy. As a substance is cooled, the KE decreases.

➡ **energy, potential energy**

large intestine

noun

The large intestine includes the colon and the rectum. No absorption of food chemicals takes place but water is removed from the undigested food.

eg The large intestine is sometimes called the lower gut.

➡ **digestive system**

laser

noun

A laser is a device for producing a narrow beam of light. It can travel over very large distances and can be focused to give enormous power intensities.

eg A laser can be used for communications, surgery and cutting.

i The word 'laser' is an acronym. The letters stand for **l**ight **a**mplification by **s**timulated **e**mission of **r**adiation.

lava

noun

Liquid rock that erupts from volcanoes is called lava.

eg Lava is extremely hot and flows quite quickly.

leaf (plural: leaves)

noun

A leaf is the main organ where food is produced by photosynthesis in plants. It usually consists of a flat green blade attached to the plant stem by a stalk.

eg This diagram shows a cross-section of a leaf.

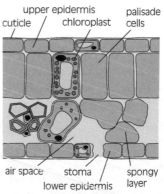

upper epidermis
cuticle / chloroplast
palisade cells

air space stoma spongy
lower epidermis layer

➡ **chloroplast, cuticle, palisade cell, spongy layer, stoma**

lever

noun

A lever is a simple machine that can be used to lift a heavy load. It consists of a rigid rod which turns or pivots about a fixed point (called the fulcrum).

eg The diagram below shows a first order lever, e.g. a seesaw or a pair of scissors.

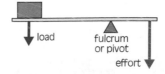

load fulcrum or pivot effort

This diagram shows a second order lever, e.g. wheelbarrow.

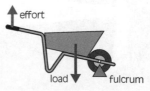

effort load fulcrum

A
B
C
D
E
F
G
H
I
J
K
L
M
N
O
P
Q
R
S
T
U
V
W
X
Y
Z

life cycle
noun

A sequence of stages through which a living organism passes, from the time of fertilisation until the same stage in the next generation, is called a life cycle.

eg Most vertebrates have a simple life cycle:

fertilisation → development as embryo → juvenile development after hatching or birth → adulthood including sexual reproduction.

Invertebrate life cycles are often more complicated.

light
noun

Light is a form of electromagnetic radiation that can be seen with the human eye. It travels freely through space. It travels in straight lines and can be reflected and refracted.

eg The speed of light is approximately 3×10^8 m/s or 186 000 miles per second.

In 1666, Isaac Newton showed that white light is made up of a mixture of lights of different colours. He shone white light through a prism.
➡ dispersion

light beam
noun

When light passes through a narrow slit, a light ray is produced. A light beam is a wide light ray.

eg A lighthouse produces a light beam that warns ships of dangerous rocks.

light beam

light gate
noun

A light gate is a beam of light. Two light gates are used together to measure time accurately. When an object breaks the first light gate a timer starts. When the object breaks the second light gate, the timer stops.

eg In the diagram, the piece of card breaks the light beam of the first light gate. Later it breaks the light beam of the second light gate.

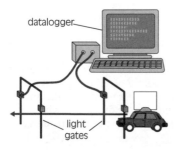
datalogger

light gates

light intensity
noun

The rate at which light is received by a unit area of a surface is called light intensity.

eg The light intensity on a sunny day is much greater than on a dull day.

limb
noun

An arm, a leg and a wing are limbs.

eg Because of their positions, limbs can easily be broken.

limestone
noun

Limestone is a sedimentary rock largely produced from the shells of dead marine animals. Much of the rock is composed of calcium carbonate, $CaCO_3$.

eg Limestone is used as a building stone and it is also used to make cement.
➡ marble, sedimentary

limitation   **? ? ?**

noun

A limitation is a restriction that is placed on something.

🔵 There is a limitation in the concentration of acids used at Key Stage 3, for safety reasons.

line graph   **? ? ?**

noun

A line graph is a graph where the points are joined by a line. This line might be straight or it might be curved. Because there may be anomalous points, the line may not go through all points.

🔵 The results of the extension of a spring when different forces are applied can be shown on a line graph.

➡ **anomalous result**

line of best fit   **? ? ?**

noun phrase

The line of best fit, which is drawn approximately in the middle of the points of a scatter diagram, enables you to estimate values not given in the original information.

🔵 The line of best fit has been drawn on the graph below.

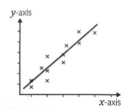

➡ **anomalous result**

liquid

noun

A liquid is a state of matter between solid and gas. A liquid has a fixed volume and takes up the shape of the bottom of the container. The particles in a liquid are close together but not regularly arranged (as in a crystalline solid). The particles are not as free to move as in a gas.

🔵 Water is a liquid at room temperature.

➡ **change of state, gas, solid**

litmus

noun

Litmus is an indicator used to test for acids and alkalis. It can be used in solution or soaked in absorbent paper such as litmus paper. Litmus turns red when added to an acid and blue when added to an alkali. It is made from a lichen that lives in the Arctic Circle.

🔵 If pieces of red and blue litmus are added to a liquid and both stay the same colour, what does this suggest about the liquid?

➡ **acid, alkali**

litre [ℓ]

noun   **? ? ?**

A litre is a metric unit of capacity. One litre is the same volume as a cubic decimetre ($1\,dm^3$) – a cube with sides 10 cm long.

🔵 How many centilitres are there in one litre?

liver

noun

The liver is a large organ found in the abdomen of humans. Its main function is to regulate the chemical composition of the blood.

🔵 The liver is one organ in the body that is able to regrow if it is damaged.

liver

loudness

noun

The ear can only make comparisons between two sounds. It cannot measure actual values. Loudness is a judgement of the level or strength of a sound reaching the ear.

eg The loudness of a sound (called intensity) can be measured using a sound-level meter. The unit of the intensity of sound is a decibel.

lubricant

noun

A lubricant is a substance used to reduce the friction between surfaces.

eg Oil is used as a lubricant. It reduces the wear on surfaces that rub together.
➡ friction

luminous (*loo*-mi-nus)

adjective

Something is said to be luminous if it gives out its own light. A non-luminous object does not.

eg The Sun and a candle flame are luminous. The Moon is non-luminous – the light from the Moon is not its own but is reflected from the Sun.

🛈 The word 'luminous' comes from the Latin word *lumen*, which means 'light'.

lung

noun

A lung is one of two spongy respiratory organs found in vertebrates. Each lung contains very large numbers of alveoli (air sacs) and blood vessels. The air breathed in is close to the blood vessels so diffusion can take place.

eg A lung is efficient because of:
- a constant supply of air
- the thinness and moisture on the surfaces
- a constant supply of fresh blood.
➡ alveolus, diffusion, respiratory system

Mm Mm Mm Mm

magma
noun

The rocks below the Earth's crust are called magma. These rocks may be molten or plastic. If the magma crystallises, igneous rocks are formed.

🟢 Magma can come to the surface through volcanic eruptions.

➡ **rock cycle**

magnetic
adjective

A magnetic material has the power to attract certain substances such as iron and steel.

🟢 Iron, cobalt and nickel are magnetic metals.

magnetic field
noun

A magnetic field is the region surrounding a permanent magnet, electromagnet or wire carrying a current where magnetic forces may be detected.

🟢 The diagram shows the magnetic field around a bar magnet.

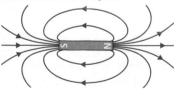

magnetic field line
noun

Diagrams that show magnetic fields, like the one above, use arrowed lines to indicate where and in which direction the magnetic field is situated. Each line is called a magnetic field line.

🟢 A magnetic field line goes from the north-seeking pole or towards the south-seeking pole.

magnitude
noun

The word 'magnitude' means size.

🟢 The magnitude of a force can be measured with a force meter.

malleable (*mal-ee-able*)
adjective

Materials that can be beaten into thin sheets are said to be malleable.

🟢 Gold is malleable and can be beaten into thin sheets called gold leaf. These leaves are so thin you can see through them.

mammal
noun

A mammal is a warm-blooded vertebrate animal. A female mammal feeds her young with milk from her breasts.

🟢 Humans and whales are each examples of a mammal.

🔒 The whale is the largest mammal known. The female whale suckles her offspring.

mammary gland
noun

A mammary gland is the milk-producing gland in a mammal.

🟢 A woman's breast and a cow's udder are examples of a mammary gland.

mantle
noun

The mantle is the part of the Earth between the crust and the core.

🟢 The mantle is made up of different types of rock.

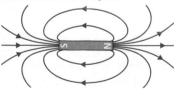

A B C D E F G H I J K L **M** N O P Q R S T U V W X Y Z

marble

noun

Marble is a metamorphic rock produced by the action of high temperature and high pressure on limestone.

eg Marble and limestone are both forms of calcium carbonate. Marble is harder than limestone and can be used for statues.

mass

noun

Mass is the quantity of matter in a body. In the SI system, the unit of mass is the kilogram. At the same place, equal masses will have the same weights as there is the same gravitational force acting upon them.

eg An object with a mass of 600 kg will have a weight of 6000 N on the Earth.
➡ **weight**

material

noun

A material is a substance out of which something is made.

eg Steel, wood and polythene are all materials. Sometimes, people restrict the word material to refer to types of cloth.

measurement ? ? ?

noun

A measurement is a size or amount found by measuring.

eg A metre rule could be used to find the measurement of the length of a table.

measuring cylinder

noun ? ? ?

A measuring cylinder can be used to measure the volume of liquids.

eg The diagram shows a measuring cylinder.

mega-

prefix ? ? ?

'Mega-' is a prefix meaning 1 000 000.

eg A megawatt is 1 000 000 watts.

meiosis (mi-o-sis)

noun

Meiosis is a type of cell division in which four daughter cells are produced. Each daughter cell contains half the number of chromosomes of the parent nucleus and results in the formation of male and female gametes.

eg The diagram shows meiosis in a cell with four chromosomes.

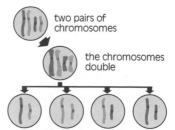

two pairs of chromosomes

the chromosomes double

four new cells are formed, each with only two chromosomes

➡ **gamete, mitosis**

melt

1 verb

To melt means to change from a solid to a liquid. This takes place at a temperature called the melting point.

eg Ice will melt at 0°C. At this temperature ice turns to water.
➡ **change of state**

2 noun

A melt is the liquid formed when a solid is melted.

eg Molten sodium chloride is a melt.

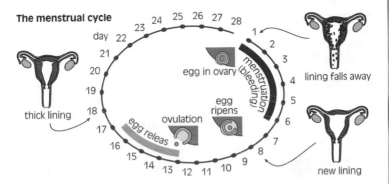

The menstrual cycle

day

menstruation (bleeding)

egg in ovary

egg ripens

ovulation

egg releases

thick lining

lining falls away

new lining

menstruation

noun

Menstruation occurs in women of child-bearing age once every 28 days if fertilisation of the ovum has not taken place. There is a discharge of blood and other fluids from the womb through the vagina.

eg Menstruation is summarised in the menstrual cycle diagram, above.

metamorphic

adjective

A rock is said to be metamorphic when it has been formed by the action of high temperatures and high pressures on existing rocks.

eg Marble is a metamorphic rock made from limestone.

➡ igneous, sedimentary

meter

noun

A meter is an instrument used for measuring.

eg Thermometer, ammeter and voltmeter are three types of measuring instrument or meter.

🔗 The word 'meter' comes from the Greek word *metron*, meaning 'a measure'.

methane [CH_4]

noun

Methane is a compound of carbon and hydrogen. It is formed inside the Earth by the action of high temperatures and pressures on plant material.

eg Methane is the main constituent of the fossil fuel natural gas.

metre [m]

noun ???

A metre is the metric unit of length equal to 100 centimetres and about 39 inches.

eg The shortest athletics event in the Olympic games is the 100-metre sprint.

micro-

prefix ???

The prefix 'micro-' means 'small'.

eg A microscope is an instrument for looking at small things.

microbe

➡ micro-organism

micro-organism

noun

A micro-organism is any tiny living organism. Micro-organisms include bacteria, viruses, protozoa, fungi and yeasts. Micro-organisms are sometimes called microbes.

eg One micro-organism might be harmful, whilst another might be beneficial. For example, micro-organisms can ferment sugar solution.

🔗 The science of micro-organisms is called microbiology.

A B C D E F G H I J K L **M** N O P Q R S T U V W X Y Z

microscope

noun

A microscope is an instrument consisting of a system of lenses that magnifies objects too small to be seen with the naked eye.

eg The diagram shows a microscope.

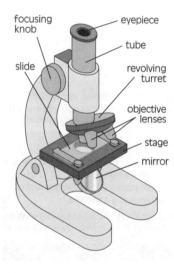

focusing knob

eyepiece

tube

slide

revolving turret

objective lenses

stage

mirror

migration

noun

The movement of animals, especially birds, from one area to another at different times of the year is called migration.

eg The migration of birds often takes place in winter when they seek warmer climates.

🛈 The word 'migration' comes from the Latin word *migrare,* which means 'to move from one place to another'.

milli- ? ? ?

prefix

The prefix 'milli-' means 'one thousandth'.

eg One milligram is the same as 0.001 g.

milligram [mg]

noun ? ? ?

A milligram is a measurement of mass equal to one thousandth of a gram.

eg A laboratory balance can weigh objects to the nearest milligram.

millilitre [ml]

noun ? ? ?

A millilitre is a measurement of capacity equal to one thousandth of a litre.

eg A millilitre is equal in volume to a cubic centimetre (1 cm^3).

millimetre [mm]

noun ? ? ?

A millimetre is a measurement of length equal to one thousandth of a metre.

eg There are ten millimetres in one centimetre.

mineral

1 noun

A mineral is an element needed in the diet in small quantities for a specific purpose.

eg Iron is required to ensure that a person is able to produce enough red blood cells. Vegetables are a good source of the mineral iron.

2 noun

Rocks from the Earth contain a number of different compounds. Each compound is called a mineral.

eg Granite contains the minerals quartz, feldspar and mica.

mirror

noun

A mirror is a polished surface that reflects light.

eg The mirror reflected the light against the ceiling.

mirror image
noun

When you look into a flat (plane) mirror, the image seen is a virtual image formed behind the mirror – but the same size as the original and upright. This is a mirror image.

eg The image produced in a plane mirror, the mirror image, is reversed.

A mirror image

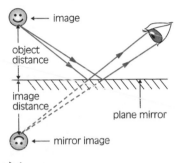

➡ **image**

mitosis
noun

Mitosis is a form of cell division in which the nucleus divides and each new cell has the same number of chromosomes as the parent cell.

eg The diagram shows mitosis in a cell with only four chromosomes.

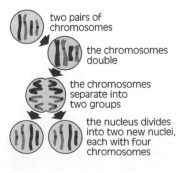

two pairs of chromosomes

the chromosomes double

the chromosomes separate into two groups

the nucleus divides into two new nuclei, each with four chromosomes

➡ **asexual reproduction, clone, meiosis**

mixture
noun

A mixture is a material consisting of two or more substances with no chemical bond. The composition of the mixture can vary. As there is no bonding between the components, the mixture can be separated by physical means.

eg A mixture of iron and sulphur can be separated using a magnet. The iron sticks to the magnet but the sulphur does not.

molecule
noun

A group of atoms in an element or compound that are chemically combined together form a molecule.

eg An oxygen molecule, O_2, has two oxygen atoms combined together. A carbon dioxide molecule, CO_2, has one carbon atom and two oxygen atoms combined in a molecule.

How many atoms are there in molecules of ozone, O_3, and methane, CH_4?

moment
➡ **moment of a force**

moment of a force
noun phrase

The moment of a force is a measure of the turning effect produced by a force. The moment of a force = force × perpendicular distance from the pivot. It is measured in units of newton metre.

eg What is the moment of the force being used to tighten the wheelnut?

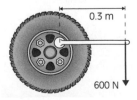

0.3 m

600 N

A
B
C
D
E
F
G
H
I
J
K
L
M
N
O
P
Q
R
S
T
U
V
W
X
Y
Z

mono-
prefix ? ? ?

The prefix 'mono-' means 'one'.

eg A monorail is a railway system where the trains run on one rail.

moss
noun

Moss is a small spore-bearing plant found in damp, shady areas.

eg During the winter, moss may spread across the lawn in the cool, damp conditions.

The word 'moss' comes from an Anglo-Saxon word, *mos*, which means 'bog'.

most appropriate equipment ? ? ?
noun phrase

When choosing equipment for an experiment, the most appropriate equipment is that which produces a consistent set of results quickly and efficiently.

eg When collecting a gas whose volume you need to measure, a burette or a gas syringe would be the most appropriate equipment.

multicellular organism
noun

A multicellular organism is an organism that is made up of many cells. An organism made of a single cell will be limited in size and complexity.

eg In multicellular organisms, different cells are specialised and have their own particular uses.

muscle
noun

Muscle is animal tissue that contracts and expands to produce movement.

eg Muscle is made of long cells that can contract to one-third of their expanded length.

➡ **antagonistic muscle**

natural gas

➡ methane

neutral
adjective
Something that is neither acid nor alkaline is said to be neutral. Neutral substances have a pH value of 7.

(eg) Water and ethanol are neutral liquids.

🔢 The word 'neutral' comes from the Latin word *neutralis*, which means 'neither'.

neutralisation
noun
Neutralisation is the process whereby exact amounts of an acid and alkali are mixed together to form a neutral solution.

(eg) A neutralisation reaction takes place when sodium hydroxide and hydrochloric acid are mixed. The word equation for the reaction is shown below.

sodium hydroxide + hydrochloric acid
$\rightarrow$ sodium chloride + water

newton [N]
noun
The newton is the SI unit of force.

(eg) One newton is the force needed to accelerate an object of mass 1 kg by 1 m/s^2.

➡ force

nickel [Ni]
noun
Nickel is a greyish-white metal used for metal alloys.

(eg) 'Silver' coins are made from an alloy of copper and nickel.

🔢 The word 'nickel' comes from a German word, *Kupfernickel*, which means 'copper devil'. It was called this by miners who confused it with copper when mining.

nitrate
noun
A nitrate is a salt produced using nitric acid.

(eg) Potassium nitrate, KNO_3, is used as a fertiliser and also to make explosives.

noise pollution
noun
Unwanted sounds are described as noise pollution. At low levels, noise pollution can make people irritable, less alert and unable to do their work efficiently. At high levels, this noise can cause temporary or permanent damage to hearing and can cause nausea.

(eg) Noise pollution can be a problem in factories and it should be monitored.

north-seeking pole
noun
When a bar magnet is hung up with a piece of string, the magnet turns until one end points towards magnetic North. This end is the north-seeking pole.

(eg) The north-seeking pole is the basis of the compass used by travellers.
➡ attraction, repulsion, south-seeking pole

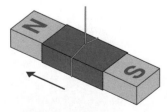

nucleus

1 noun

The nucleus of a cell is the inner part of the cell which contains the genetic information.

eg The nucleus of a cell controls the actions of the cell.

➡ **cell**

2 noun

The nucleus of an atom is the central part of the atom.

eg The nucleus of a carbon atom is made up of protons and neutrons.

i If an atom was magnified to be the size of a football stadium, the nucleus would be the size of a pea on the centre spot.

nutrient

noun

A nutrient is any chemical required by an organism to live, grow and reproduce.

eg Plants need only simple nutrients such as nitrate and can make complex materials such as amino acids and proteins. Animals need more complex nutrients.

i The word 'nutrient' comes from the Latin word *nutrire*, which means 'to nourish'.

nutrition

noun

Nutrition is the process by which an organism obtains the nutrients it needs to live, grow and reproduce. Green plants make their own food by photosynthesis; other organisms are consumers.

eg There are people who specialise in studying nutrition, called nutritionists.

Oo

observation   ? ? ?

noun

An observation is something that has been detected by one of the senses and recorded.

eg Be careful to make your observations carefully.

obsidian

noun

Obsidian is a volcanic glass formed by the rapid cooling of granite magma.

eg Obsidian is usually black but can be red or brown in colour.

opaque (*o-payk*)

adjective

A substance is opaque if light cannot pass through it. The opposite of opaque is transparent.

eg A block of wood is opaque and a piece of glass is transparent.

i The word 'opaque' comes from the Latin word *opacus*, which means 'dark' or 'shaded'.

➡ transparent

opinion   ? ? ?

noun

An opinion is a belief that seems likely but that is not based upon proof.

eg In the opinion of many scientists, recent events involving flooding and drought are caused by global warming.

i The word 'opinion' comes from the Latin word *opino*, meaning 'belief'.

orbit

1 noun

The path in space of one celestial body around another is called an orbit.

eg The orbit of Mercury around the Sun is an ellipse.

2 verb

To orbit means to move around a celestial body in a curved path.

eg Venus orbits the Sun with an almost circular orbit. Can you name any other such planets?

order of reactivity
➡ reactivity series

organ

noun

An organ is a part of a body or plant which has a special function, e.g. liver or kidney.

eg The diagram shows some of the organs in a human.

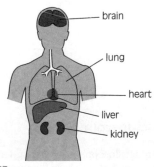

- brain
- lung
- heart
- liver
- kidney

i The word 'organ' comes from the Latin word *organum*, which means 'instrument'.

➡ brain, heart, kidney, liver, lung, skin

organism
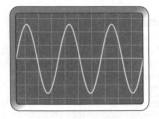

noun

An organism is any plant, animal, fungus or bacterium.

eg It is believed that three-quarters of tiny organisms, such as insects and spiders, have never been positively identified.

oscillation (*o-sill-ay-shun*)

noun

One complete to-and-fro movement of a vibrating object or system is called an oscillation. The time for one oscillation is called the period, and the number of oscillations each second is called the frequency.

eg A swinging pendulum moves from the central position to the left, back to its central position, to the right and back to its central position. This is one oscillation.

maximum potential energy, zero kinetic energy

losing potential energy, gaining kinetic energy

zero potential energy, maximum kinetic energy

gaining potential energy, losing kinetic energy

oscilloscope

noun

An oscilloscope is an instrument used to measure how electrical voltages or potentials change over time. The results are displayed as a signal.

eg The signal on the oscilloscope is displayed on a cathode ray tube.

The diagram shows an oscilloscope trace. How many waves are shown?

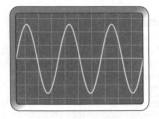

ovary

noun

The organ where the ovum (egg cell) is produced in female animals is called the ovary.

eg A woman has two ovaries. Each ovary is located in the abdomen near the end of the fallopian tubes.

➡ **reproductive systems**

overwintering

noun

Dahlia tubers, for example, are lifted in the autumn and stored in a frost-free place, before being replanted in spring. This is called overwintering.

eg Overwintering protects plants from frosts.

oviduct

noun

The tube through which the ovum travels from the ovary to the uterus is called the oviduct.

eg If a woman's oviduct is blocked, she could be infertile.

ovulation

noun

Ovulation is the process of making and releasing the ovum (egg cell).

eg Ovulation takes place every month in a woman.

➡ **menstruation**

ovum (plural: ova)

noun

An unfertilised female egg cell is called an ovum.

eg Female mammals all produce at least one ovum.

➡ egg, gamete, sperm, zygote

oxide

noun

An oxide is a compound formed when an element reacts with oxygen. Sometimes this reaction involves burning. Oxides of metals are generally neutral or alkaline. Oxides of non-metals are acidic.

eg Magnesium oxide is formed when magnesium burns in oxygen.

oxygen [O]

noun

Oxygen is a colourless, odourless gas which is very reactive. It makes up about 20% of the Earth's atmosphere. Oxygen is required for respiration, combustion and corrosion of metals, and is produced during photosynthesis. Oxygen can exist in two forms (called allotropes). Oxygen usually exists with molecules containing pairs of atoms (O_2). It can also exist in a reactive form as ozone (O_3) with three oxygen atoms in each molecule.

eg Oxygen can be tested for with a glowing splint. When the splint is put into oxygen, the splint relights.

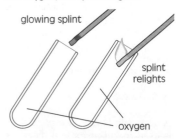

glowing splint

splint relights

oxygen

Joseph Priestley is credited with discovering oxygen in 1774. However, it may, in fact, have been discovered a couple of years earlier by Carl Scheele but his experiments were not documented.

oxygen concentration

noun

The oxygen concentration is the amount of oxygen in a given volume of gas.

eg At high altitudes above the Earth, there is a very low oxygen concentration. Climbers need to carry oxygen tanks to help them breathe.

ozone

noun

Ozone is a colourless gas with a chlorine-like smell, that is found in the upper layers of the Earth's atmosphere.

eg Ozone screens out harmful ultraviolet radiation from the Sun by acting as a barrier.

ozone depletion

noun

The use of chemicals such as chlorofluorocarbons (in refrigerators and aerosols) has led to holes in the ozone layer around the Earth. This reduction in ozone levels in the atmosphere is called ozone depletion.

eg In order to reduce ozone depletion, harmful chemicals have been replaced.

A B C D E F G H I J K L M N O P Q R S T U V W X Y Z

palisade cell

noun

A palisade cell is a cell found near the surface of leaves. It contains a large number of chloroplasts.

eg Since they contain a large number of chloroplasts, the palisade cells are where photosynthesis takes place.

➜ **chloroplast, cuticle, leaf, spongy layer, stoma**

parallel

1 adjective

In a parallel circuit, the current splits at a junction and then rejoins.

eg The diagram shows a parallel circuit. What happens when one lamp fails?

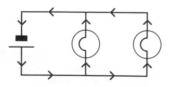

2 adjective ? ? ?

Lines that are parallel always stay the same distance apart and never meet.

eg The diagram shows two parallel lines.

particle

noun

A tiny unit of matter is called a particle. The word 'particle' is used when the exact nature of the particle is not known or is not necessary. Particles can be atoms, molecules or ions.

eg Oxygen gas is made up of particles. These particles are molecules of oxygen with two oxygen atoms in each molecule.

particle theory

noun

An understanding of the number, position and movement of particles leads to particle theory.

eg Particle theory is used to explain many scientific observations.

pathogen

noun

A pathogen is a micro-organism such as a bacteria or virus that causes a disease in an organism.

eg Most pathogens do not deliberately cause disease. The disease is incidental in the pathogen's search for food and shelter.

pattern ? ? ?

noun

When you can see that results show a trend, there is a pattern in the results.

eg A break in a pattern of results can indicate a problem with an investigation.

penis

noun

The penis is the male reproductive organ which transfers sperm into the vagina of the female.

eg The tube which carries urine and sperm through the penis is called the urethra.

➜ **reproductive systems**

pesticide

noun

A pesticide is any chemical that is used to kill pests and diseases.

There are three main types of pesticides: insecticides, herbicides and fungicides.

eg Many pesticides that were once used are no longer used, as they are too toxic.

➡ **fungicide, herbicide, insecticide, weedkiller**

petal

noun

The petal is part of a flower that attracts insects or birds to pollinate the plant.

eg Petals are often large, brightly-coloured and scented to attract insects or birds. The diagram shows the parts of a flower.

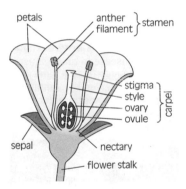

pH (pH range)

noun

The pH range is a scale from 0–14 which shows how acidic or alkaline a substance is. A pH value of 7 is neutral. Below 7 is acidic and the lower the number, the stronger the acid. Above 7 is alkaline and the higher the number, the stronger the alkali.

eg The pH of something can be found using universal indicator or using a pH meter.

phloem *[flome]*

noun

Phloem is a type of plant tissue.

eg Phloem transports sugars and other nutrients from the leaves to other parts of a plant.

➡ **xylem**

phlogiston *(flo-jis-ton)*

noun

In the 17th century, scientists tried to explain combustion. They believed that when a substance burned, it gave out a substance called phlogiston. It was largely because of the work of Lavoisier that the modern ideas of combustion developed.

eg It was difficult to match the gain in mass on combustion with the loss of phlogiston.

photosynthesis

noun

Photosynthesis is the process that takes place in green plants when carbon dioxide and water vapour are converted into glucose and oxygen. Light is needed for the process to take place.

eg The process of photosynthesis takes place in the presence of the green pigment, chlorophyll.

pitch

noun

The pitch of a sound is how high or low the sound is.

eg The diagrams show the oscilloscope traces for two sounds. The one with a higher pitch (right) has a greater frequency – that is, more waves in a given time.

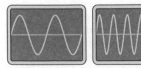

pivot

1 noun

A pivot is a central point around which something revolves, turns, balances or oscillates.

eg The pivot is at the centre point of the seesaw.

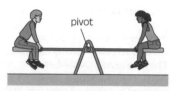

pivot

2 verb

To pivot means to revolve, turn or balance as if on a pivot.

eg The seesaw pivots at its centre.

placenta

noun

The placenta in mammals is a disc-shaped organ attached to the uterus during pregnancy. The placenta provides the embryo with nutrients and oxygen, and takes away waste products.

eg When the baby is born, the placenta is discharged from the mother's womb.

planet

noun

A planet is a large celestial body that orbits around a star.

eg The planets in our Solar System are Mercury, Venus, Earth, Mars, Jupiter, Saturn, Uranus, Neptune and Pluto.

plant food

➡ nutrient

plaster of Paris

noun

Plaster of Paris is a white powder that sets solid when mixed with water. It is used for plastering walls and for making plasterboard. It is also used to make plaster casts for broken bones.

eg Plaster of Paris is a hydrate of calcium sulphate, $2CaSO_4.H_2O$.

ℹ Plaster of Paris was first given this name because it was widely used in Paris as there were deposits of calcium sulphate nearby.

plastic

adjective

Materials that can be softened and moulded by heat and/or pressure are said to be plastic.

eg When polystyrene is heated, it becomes plastic and can be made into ceiling tiles.

ℹ The word 'plastic' comes from the Greek word *plastikos*, meaning 'moulded'.

pneumatic (*new-ma-tic*)

adjective

A tool or piece of machinery operated or driven by compressed air is said to be pneumatic.

eg A pneumatic drill is usually used to break up the surface of a road.

pollen

noun

Pollen is a dust-like powder produced by the anthers of flowering plants.

🔵 A gardener often uses a paintbrush to transfer pollen from one plant to another.

🔴 The word 'pollen' comes from the Latin word meaning 'fine dust'.

pollination (verb: pollinate)
noun

The process of transferring pollen from the anther to the stigma is called pollination.

🔵 Pollination leads to fertilisation and the subsequent formation of seeds.

poly- ???
prefix

The prefix 'poly-' means 'many'.

🔵 Polystyrene is a plastic made of many styrene units.

population size
noun

The population size is the number of things living within a given area.

🔵 The city's population size kept growing.

porous
adjective

A porous material is one that has tiny holes or pores within it. It means that liquids or gases can pass through.

🔵 A sandy soil is a porous soil. There are gaps in the soil that air and water can pass through.

🔴 The word 'porous' comes from the Latin word *porosus*, which comes from *porus*, meaning 'a pore'.

potential difference [pd]
noun

Potential difference (voltage) is the difference in electrode potential from one point to another. In a conductor, the difference in potential will cause an electric current to pass. The unit of potential difference is the volt (V).

🔵 The current passing through a 3Ω resistor is 0.5 A. What is the potential difference across the resistor? Use the relationship $V = IR$.

potential energy
noun

Potential energy is the energy possessed by an object because of its vertical position or its state.

🔵 Water at the top of a waterfall possesses potential energy because of its position. As the water falls down, this potential energy is converted into kinetic energy. A stretched spring or elastic band also possesses potential energy.

🔴 James Joule, while on honeymoon with his wife Amelia, measured accurately the temperature at the top and bottom of a waterfall. He worked out that the water at the bottom of the waterfall would be at a slightly higher temperature as it has given up its potential energy.

➡ **energy, kinetic energy**

precipitation
noun

A reaction where a solid (called a precipitate) is formed in a solution is called precipitation.

🔵 When barium nitrate solution and sulphuric acid are mixed, precipitation occurs and a white precipitate of barium sulphate is formed.

barium nitrate + sulphuric acid →
barium sulphate + nitric acid

🔴 You may use this word differently in Geography. Here it is used to describe water that falls from clouds in the form of rain, snow, hail, etc.

precision ???
noun

When a piece of apparatus is used with precision, it is used as accurately as possible.

🔵 When a burette is used with precision, it can measure a liquid to the nearest $0.05 \, cm^3$.

predator
noun

A predator is any animal that gets its food by catching and eating other animals.

eg A lion catches an antelope for food. The lion is the predator and the antelope is the prey.

i The word 'predator' is a 20th-century word. It comes from the Latin word *praedator*, which means 'to plunder'.

➜ **prey**

prediction (verb: predict)
noun ? ? ?

When carrying out an investigation, you should try to make some statement about what you expect the investigation to uncover. This is called a prediction. You should also try to use your scientific knowledge to justify your prediction.

eg When dissolving different amounts of sugar in the same amount of water at the same temperature, you might make the prediction that the larger the amount of sugar used, the longer it will take to dissolve. You might try to use ideas of particles to explain your prediction.

present
1 verb (*pre-sent*) ? ? ?

When you present something, you explain it to someone else. This may be done at a presentation.

eg The children were asked to present their results to the teacher.

2 adjective (*pres-ent*)
? ? ?

In routine tests, if a substance being tested for is there, it is said to be present.

eg In a starch test, iodine is added to a food. If the colour of the iodine turns from brown to dark blue-black, this shows that starch is present. If it does not change colour, starch is not present – it is absent.

pressure
noun

Pressure is the force acting on a unit area of a surface. It is calculated using the formula:

$$pressure = force \div area$$

It is measured in units of newtons per square metre (N/m^2) or pascals (Pa).

eg A bulldozer weighs 150 000 N. The area of its two tracks in contact with the ground is $10 m^2$. The pressure is, therefore, $150 000 \div 10 = 15 000$ Pa.

prey
noun

An animal that is caught by a predator is called the prey.

eg A lion catches an antelope for food. The lion is the predator and the antelope is the prey.

➜ **predator**

primary colour
noun

A primary colour is one of three colours (usually red, green and blue) that can be mixed to make any other colour, including white but excluding black.

eg On a television screen, the three primary colours combine to produce white light.

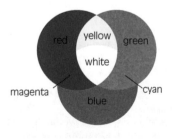

➜ **secondary colour**

prism
noun

A prism is a transparent, triangular block.

eg When white light is passed through a prism, it can be split up into the colours of the spectrum. Which colours make up the spectrum?

➡ **dispersion**

producer

noun

A producer is an organism that produces its own food.

eg A producer is always a plant. It can make its food by the process of photosynthesis.

product

noun

A substance that is made during a chemical reaction is called a product.

eg Magnesium and sulphuric acid react together to form magnesium sulphate and hydrogen. Which substances are the products?

propagation

noun

Growing new plants is called propagation.

eg Propagation takes place when a gardener takes cuttings of plants to get a supply of new plants.

property (plural: properties)

noun

A characteristic of a material is called a property.

eg Melting point, density, hardness and colour are all properties of a material. The property may be chemical or physical.

proportional [$\propto$]   **? ? ?**

adjective

When you draw a straight line graph and the graph passes through the origin (0,0), this shows that one variable is proportional to the other.

eg The diagram shows a graph of current against voltage. As the graph is a straight line and it goes through the origin, voltage is proportional to current.

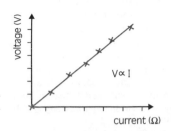

$V \propto I$

protease (*pro-tee-aze*)

noun

A protease is an enzyme that is capable of digesting proteins.

eg Enzyme washing powders contain a protease that will digest protein stains such as blood.

protein

noun

Protein is the essential chemical required for the growth and repair of the body. Proteins are found in soya beans, fish, meat and cheese. Proteins are long-chain molecules made up of amino acid molecules joined together.

eg Proteins such as keratin and collagen make up skin and bone.

proximity

noun

The word proximity means being close in space.

eg The meteorite was in close proximity to the Earth.

puberty

noun

The onset of sexual maturity in humans and other primates is called puberty.

eg Puberty is the stage during which the bodies of young people change and they become adults.

➡ **adolescence**

A B C D E F G H I J K L M N O P Q R S T U V W X Y Z

pulse

noun

An impulse transmitted through the arterial system is called a pulse. When the heart muscle contracts, blood is pumped from the heart into the aorta – the large artery leading out of the left ventricle. Because the arteries are elastic, the sudden rise in pressure causes a throb through them.

eg Everybody alive has a pulse.

pulse rate

noun

The pulse rate is the number of pulses each minute. On exercising, the pulse rate rises but it then returns to the original rate on resting. The pulse rate can be felt anywhere where an artery comes near to the surface, for example in the wrist or the neck.

eg In a human, the pulse rate at rest is about 70 beats per minute.

pumice

noun

Pumice (sometimes called pumice stone) is a light grey or white solidified lava. The bubbles within the structure are caused by gas escaping through the rock as it solidifies.

eg Pumice is used as an abrasive.

pump

noun

A pump is a device for forcing or driving gases or liquids in or out of a container.

eg A pump in a central heating system forces the water round the pipes to reach each radiator.

pyramid of numbers

noun

A diagram that shows the numbers of different organisms at different trophic levels is called a pyramid of numbers.

eg The diagram below shows a pyramid of numbers for a food chain based on an oak tree.

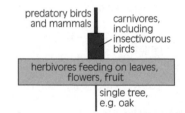

predatory birds and mammals | carnivores, including insectivorous birds

herbivores feeding on leaves, flowers, fruit

single tree, e.g. oak

quadrat sampling

noun

A quadrat is a square structure used for counting the number of plants or animals within a given area. Quadrats are often 0.5 m or 1 m square. The quadrat is placed on the ground and the number of organisms within that area are counted. By taking sufficient representative samples, a good estimate can be made of the number present in a larger area. This is called quadrat sampling.

eg The number of daisies on the school field can be found by quadrat sampling.

qualitative

adjective

A study is said to be qualitative when it involves looking only at changes that do not involve quantities.

eg Testing the change that takes place when copper sulphate crystals are heated is a qualitative exercise.

quantitative

adjective

A study is said to be quantitative when it involves quantities.

eg Finding out the mass of copper produced from a given mass of copper oxide is a quantitative exercise.

quartz

noun

Quartz is a colourless mineral made of silicon dioxide.

eg Some gemstones are quartz coloured with impurities.

quartzite

noun

Quartzite is a hard, durable rock composed largely or entirely of quartz.

eg Quartzite is also a sandstone in which the grains are cemented together with silicon dioxide.

radiation

noun

Radiation is the emission of energy from a source. The energy travels as waves or particles through a medium (air or a vacuum).

eg Radio waves, microwaves, infrared waves, visible light, alpha particles and beta particles are all forms of radiation.

rainbow

noun

A rainbow is a bow-shaped display in the sky of the colours of the spectrum. It is caused when light passes through raindrops and dispersion takes place.

eg What are the seven colours of the rainbow?

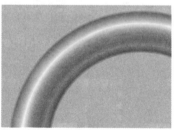

➡ **dispersion, spectrum**

range

noun

The range is the statement of the limits within which any changes can take place in an investigation.

eg The temperature range studied was between 20°C and 60°C.

reactant

noun

A substance that reacts in a chemical reaction is called a reactant.

eg When calcium carbonate reacts with hydrochloric acid, calcium chloride, water and carbon dioxide are formed. What are the reactants?

reaction

noun

A chemical change in which reacting substances (reactants) are changed into products is called a reaction. The reaction can be summarised either by a word or symbol equation.

eg The reaction of sodium and chlorine produces sodium chloride.

➡ **product, reactant, symbol equation, word equation**

reactivity series

noun

The reactivity series is a list of metals in order of their reactivity. The most reactive metal is at the top of the list and the least reactive at the bottom.

eg The reactivity series you often see is:

● Potassium
● Sodium
● Calcium
● Magnesium
● Aluminium
● Zinc
● Iron
● Lead
● Copper
● Silver.

Other metals could also be included. You do not need to remember this order. It will be given to you when you need it.

record *(ree-cord)*

verb

To record something is to make a note of it, usually in writing.

(eg) When you take some results it is important to write them down, or record them straight away.

red blood cell

noun

Red blood cells contain the red pigment haemoglobin. They transport oxygen around the body. The cells are disc-shaped to ensure they have a large surface area compared with their volume. Red blood cells have no nucleus.

(eg) Red blood cells can screw themselves up to pass through the narrowest capillaries.

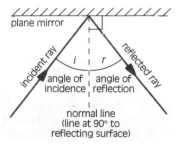

reflection (verb: reflect)

noun

The turning back of a light ray when it hits a smooth surface that it does not enter (e.g. a mirror) is called a reflection.

(eg) The diagram shows the reflection of light in a plane mirror.

angle of incidence i | angle of reflection r

normal line
(line at 90° to reflecting surface)

refraction

noun

Refraction is the bending of light when it passes from one medium to another. Refraction takes

place because light waves travel at different velocities in different media. For example, the speed of light in water and glass is 230 000 000 m/s and 200 000 000 m/s respectively.

(eg) The diagram shows the refraction of light through a piece of glass.

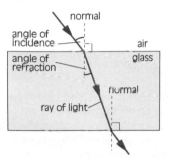

normal

angle of incidence

air

glass

angle of refraction

normal

ray of light

relay

noun

A relay is a switch operated by turning an electromagnet on and off. There is a small current in the relay circuit which switches on a large current in the main circuit.

(eg) The diagram shows a relay being used to switch a mains lamp on and off.

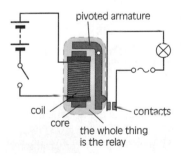

pivoted armature

coil

core

contacts

the whole thing is the relay

reliability   ? ? ?

noun

Reliability in your results means you have confidence that if your experiment were repeated by somebody else, their results would be similar to yours.

eg You can get a good idea of the reliability of your results by looking at the results of other people in your class.
➡ **evaluation**

reliable data ? ? ?
noun

Reliable data are data on which you can rely, or that you can trust to be true.

eg The class produced reliable data from their investigation.

repeat ? ? ?
verb

To repeat something means to do something more than once.

eg The experiment was repeated several times to make sure it was accurate.

repeat measurement
noun ? ? ?

A repeat measurement is a measurement that you do again. It is only worth repeating a measurement if there could be some doubt about it.

eg If you are timing ten swings of a pendulum, you might carry out repeat measurements and work out an average of your results. It should not be necessary to repeat measurements when measuring the length of a table.

repeat observation
noun ? ? ?

If there is doubt about an observation during an experiment, you should do a repeat observation.

eg A repeat observation allows you to check the observation you made previously.

repeat reading ? ? ?
noun

A repeat reading is a second reading taken in an investigation. When carrying out an investigation, it is wise to plan to repeat readings (observations and measurements). You should average your readings but exclude any that are very different. If the readings you take are very similar, it shows that there is reliability.

eg In an investigation, repeat readings are taken. These are $25.0\,cm^3$, $24.8\,cm^3$, $31.2\,cm^3$ and $24.9\,cm^3$. What average should you take from these readings?

reproduction
(verb: reproduce)
noun

Reproduction is the process whereby a living organism produces another organism similar to itself. Reproduction may be sexual or asexual.

eg Asexual reproduction produces identical organisms but sexual reproduction does not.
➡ **asexual reproduction, sexual reproduction**

reproductive systems
plural noun

The reproductive systems are the organs in females and males needed for reproduction.

eg The diagrams show the reproductive systems in a human female and a human male.

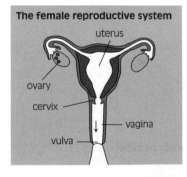

The female reproductive system

uterus
ovary
cervix
vagina
vulva

The male reproductive system

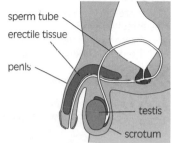

sperm tube
erectile tissue
penis
testis
scrotum

➡ **ovary, penis, testis, uterus, vagina**

reptile
noun

A reptile is a cool-blooded vertebrate with a scaly body.

eg Examples of reptiles are lizards, snakes, tortoises and crocodiles.

🔢 The word 'reptile' comes from the Latin word *repere*, meaning 'to creep or crawl'.
➡ **vertebrate**

repulsion
noun

Repulsion is a force that tends to push two objects apart.

eg When the north poles of two bar magnets are brought close together, repulsion occurs.
➡ **attraction**

resistance [R]
1 noun

The opposition of a circuit component to the flow of charge is called resistance. The SI unit of resistance (R) is the ohm (Ω).

eg The resistance is the potential difference (voltage divided by the current). Calculate the resistance of a component where $V = 6\,V$ and $I = 2\,A$.

2 noun

Resistance (drag) is the frictional force opposing the motion of an object through a liquid or gas.

eg A racing cyclist is streamlined to reduce resistance.

respiration (verb: respire)
noun

Respiration is the process occurring in cells by which food compounds react with oxygen to produce carbon dioxide and water. The word equation is:

glucose + oxygen →
carbon dioxide + water

eg The process of respiration releases energy.

🔢 The word 'respiration' comes from the Latin word *respirare*, which means 'to breathe'.

respiratory system
noun

All the organs that allow mammals to breathe are together called the respiratory system.

eg The respiratory system consists of the trachea, bronchioles, lungs and diaphragm, as well as other organs.
➡ **breathing**

A B C D E F G H I J K L M N O P Q R S T U V W X Y Z

A B C D E F G H I J K L M N O P Q R S T U V W X Y Z

result ? ? ?

noun

A result is the outcome of something.

eg Results are obtained during an investigation.

reversible

adjective

Something is said to be reversible when it is capable of being reversed or returned to an original condition. A reversible reaction is a reaction that goes both ways. The reactants can be converted to the products and the products can return to the reactants.

eg When blue copper sulphate crystals are heated, water and anhydrous copper sulphate are formed. If water is added to anhydrous copper sulphate, blue copper sulphate is reformed. Therefore, the reaction is reversible.

The word 'reversible' comes from the Latin word *revertere*, which means 'to turn back'.

revolve

verb

To revolve means to turn or rotate.

eg The minute hand of an analogue clock revolves once each hour.

ribcage

noun

There are curved bones called ribs attached to the spine of a vertebrate that protect the lungs and the heart. These ribs make up the ribcage.

eg In a human, there are 12 pairs of ribs in the ribcage.

rigid (noun: rigidity)

adjective

A material is said to be rigid if it is completely stiff and inflexible.

eg A bar of steel is a rigid material.

The word 'rigid' comes from the Latin word *rigere*, which means 'to be stiff'.

rock cycle

noun

The rock cycle summarises the processes whereby new rocks are formed and other rocks gradually return to magma.

eg The rock cycle is driven by the energy from the Sun and nuclear decay within the Earth (see below).

rotation (verb: rotate)

noun

The act of rotating, or spinning is called rotation.

eg The rotation of the Earth can be measured.

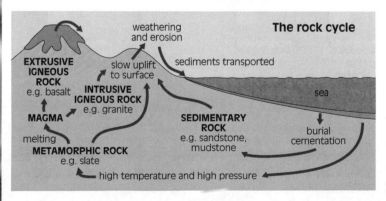

The rock cycle

weathering and erosion

EXTRUSIVE IGNEOUS ROCK
e.g. basalt

slow uplift to surface

sediments transported

INTRUSIVE IGNEOUS ROCK
e.g. granite

MAGMA

sea

melting

METAMORPHIC ROCK
e.g. slate

SEDIMENTARY ROCK
e.g. sandstone, mudstone

burial cementation

high temperature and high pressure

salt

noun

A salt is a compound formed when the hydrogen of an acid is replaced by a metal. For example, replacing the hydrogen atoms in sulphuric acid produces the salt sodium sulphate:

$$H_2SO_4 \rightarrow Na_2SO_4$$

Salts can be produced by the action of metals, metal oxide, metal hydroxide and metal carbonates on the appropriate acid.

eg The word 'salt' is often used for common salt, sodium chloride, NaCl.

sample size   ? ? ?

noun

The number of objects chosen for an investigation is called the sample size.

eg It is important that the sample size chosen is enough to establish any true pattern but not too large or it may be unmanageable.

sampling   ? ? ?

noun

The process of choosing and collecting a sample is called sampling.

eg After sampling from the local pond, students could count the number of different organisms.

sandstone

noun

Sandstone is a sedimentary rock made of sand grains cemented together with clay.

eg Sandstone is used as building material.

satellite

noun

A celestial body that orbits a much larger celestial body is called a satellite.

eg The Moon is a satellite of the Earth.

i The word 'satellite' comes from the Latin word satelles, which means 'attendant'. The first artificial satellite, Sputnik 1, was launched into orbit around the Earth in 1957.

saturated solution

noun

A saturated solution is a solution in which the maximum mass of a solid, liquid or gas is dissolved at a given temperature.

eg If more salt is added to a saturated salt solution at room temperature, the additional salt will stay undissolved at the bottom of the solution.

➡ solubility

scatter graph   ? ? ?

noun

A graph that shows the relationship between two quantities, such as height and weight, without joining the points in a line, is called a scatter graph.

eg The scatter graph showed a clear pattern in the results.

scientific method ???

noun

Scientists use a scientific method when trying to explain observations that they make. They look for patterns in the data and they attempt to come up with a hypothesis which they can test by experiment. If the experiments show this hypothesis to be true, it is then called a theory.

eg You will learn about scientific method when you study science.

➡ **hypothesis, theory**

scrotum

➡ **testis**

secondary colour

noun

Mixing two primary colours produces a secondary colour.

eg The table gives the secondary colour formed when different primary colours are mixed.

primary colours mixed		secondary colour produced
red	green	yellow
red	blue	magenta
green	blue	cyan

➡ **primary colour**

secondary data ???

noun

Secondary data are data you may download from sources such as the Internet or get from books. They consist of information that has been gathered by someone else that supports or reinforces your own data.

eg Secondary data are often useful in planning your investigation. They may give you an indication of the range of results you need to take.

sedimentary

adjective

The word 'sedimentary' describes a kind of rock that is produced when fragments of rock (sediments) are deposited, compressed and compacted together. Sedimentary rocks are also produced when solutions evaporate, e.g. halite or sodium chloride.

eg Limestone, sandstone and shale are sedimentary rocks.

 Two-thirds of the Earth's surface is made up of sedimentary rocks.

➡ **igneous, metamorphic**

selective breeding

noun

A way of improving stock by selecting and breeding from those animals and plants that have desired characteristics is called selective breeding.

eg Beef cattle can be bred for more meat by choosing parents with the desired characteristics. After a few years of selective breeding, the cattle will develop the desired characteristics.

sepal

noun

The sepal the green part of a plant that covers and protects the flower in the bud.

eg The sepal is a modified leaf.

separate

verb

To separate means to force or keep things apart.

eg There are different methods used to separate mixtures. The method chosen depends upon the properties of the substances being separated.

sequence of events ???

noun phrase

When planning an investigation, it is important to get the various steps in the correct order. This is called the correct sequence of events.

(eg) In order to check you have the correct sequence of events, read your plan through and check that you have everything in the correct order.

series

adjective

Series means side by side. In a series circuit, the electricity passes through each component in turn.

(eg) The diagram shows a series circuit. What happens when one lamp fails?

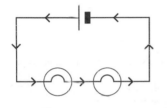

sex cell
➡ gamete

sexual reproduction

noun

Sexual reproduction is a process that requires the union or fertilisation of two gametes (an egg cell and a sperm cell) to form a zygote.

(eg) Sexual reproduction produces variation – the offspring are not identical to the parents.

shadow

noun

A shadow is a dark image or shape that is formed when light cannot pass through an opaque object. It is an area where light cannot reach.

(eg) A solar eclipse is caused when light from the Sun is blocked out by the Moon and a shadow covers the Earth.

shale

noun

Shale is a fine-grained sedimentary rock formed by the compression of clay, sand or silt.

(eg) Shale is converted into slate by high temperatures and high pressures.

SI and SI unit

???

abbreviation

SI is the name for a modern system of units used in the measurement of all physical quantities.

(eg) The SI system is based on seven basic units: the metre (m) for length, the kilogram (kg) for mass, the second (s) for time, the ampere (A) for electric current, the Kelvin (K) for temperature, the candela (cd) for luminous intensity and the mole (mol) for the amount of substance.

(i) The letters 'SI' come from Système International d'Unites.

side effect

noun

When taking a medicine, there may be another thing that happens apart from the desired effect. This is called a side effect.

(eg) One side effect of this medicine is that it causes drowsiness.

signal generator

???

noun

A signal generator is a device that turns electrical energy into sound.

(eg) A signal generator can be used to generate a wide range of sounds.

skin

noun

Skin is a tough, flexible, waterproof covering of a human or animal body.

(eg) Skin is the first line of defence preventing bacteria entering the body.

slate
noun
Slate is a dark-grey metamorphic rock formed by the action of high temperature and high pressure on clays and shales.

eg Slate can be split into sheets and used for roofing.

small intestine
noun
The small intestine is the part of the digestive system between the stomach and the large intestine. In the small intestine, food chemicals such as glucose are absorbed into the bloodstream.

eg Absorption takes place efficiently in the small intestine because there are a large number of villi which give a large surface area.

➡ **digestive system**

sodium [Na]
noun
Sodium is a soft, silvery-white reactive metal.

eg Sodium reacts rapidly with cold water.

Solar System
noun
The Solar System is the collective name for the Sun and all the celestial bodies that orbit around it.

eg The edge of the Solar System extends to about 1.5 light years – about one-third of the way towards the nearest star.

solenoid
noun
A solenoid is a coil of wire that produces a magnetic field when an electric current is passed through it.

eg When an iron core is placed inside a solenoid, an electromagnet is formed.

solid (verb: solidify)
noun
A solid is a substance in a physical state that resists changes to its size and shape. In a solid, the particles are closely packed together and are only able to vibrate.

eg When molten sulphur is left to cool, it solidifies and the yellow solid sulphur is formed.

➡ **change of state, gas, liquid**

solubility
noun
The solubility of a substance is the number of grams of the substance that can dissolve in 100 g of the solvent at a given temperature.

eg The solubility of salt in water at 30°C is 36 g per 100 g of water.

➡ **dissolve, solute, solution, solvent**

soluble
adjective
A substance is said to be soluble when it dissolves in a liquid.

eg Salt is soluble in water.

solute
noun
The substance that is dissolved in a solvent is called the solute.

eg When salt is dissolved in water, the salt is the solute.

➡ **dissolve, solubility, solution, solvent**

solution
noun
The mixture produced when a solute is completely dissolved in a solvent is called a solution.

eg When salt is dissolved in water, a salt solution is formed. Can something that has been dissolved be recovered?

➡ **dissolve, solubility, solute, solvent**

solvent

noun

A solvent is a liquid in which another substance can be dissolved.

eg Water is a solvent for the widest range of substances.

➡ **dissolve, solubility, solute, solution**

sound

noun

Sound is made when something vibrates in a to-and-fro motion. These vibrations need a medium to transfer through. They move fastest through solids and slowest through gases.

eg Televisions, radios and hi-fis all use loudspeakers that vibrate and produce sound.

south-seeking pole

noun

When a bar magnet is hung up with a piece of string, the magnet turns until one end points north. At the other end of the magnet is the south-seeking pole.

eg A south-seeking pole is attracted to a north-seeking pole.

➡ **attraction, north-seeking pole, repulsion**

species (plural: species)

noun

A species is a group of organisms that resemble each other and can interbreed to produce fertile offspring.

eg All human beings belong to the same species. They may differ in many ways but they can interbreed.

spectrum (plural: spectra)

noun

The spectrum is the arrangement of different colours of visible light in order of frequency or wavelength.

eg The colours of the spectrum are red, orange, yellow, green, blue, indigo and violet.

🛈 You can remember the colours of the spectrum by remembering the name ROY G BIV.

➡ **dispersion**

sperm

noun

The sperm or sperm cell is the male gamete.

eg A sperm cell can move great distances before internal fertilisation takes place.

🛈 The word 'sperm' comes from the Greek word *sperma*, which means 'seed'.

➡ **gamete, ovum, zygote**

sphere (adjective: spherical)

noun

A sphere is a circular solid shape in which all points on the surface are equidistant from the centre.

eg The ball bearing is a perfect sphere.

spin

verb

To spin means to rotate quickly.

eg A spinning top continues to spin for a long time.

spongy layer

noun

The spongy layer is a layer near the bottom of a leaf, consisting of large cells that do not pack closely together.

eg There are air spaces in the spongy layer where oxygen and carbon dioxide can be stored.

➡ **leaf**

stamen

noun

The stamen is the main male reproductive organ of a flower.

eg The pollen grains are produced in the stamen.

➡ **petal**

starch

noun

Starch is a carbohydrate made up of many glucose units joined together. It is produced by plants after photosynthesis and stored in the roots, tubers, seeds and fruit. Starch can be tested for, using iodine, which turns a dark blue-black colour.

eg Starch is split up with enzymes or hydrochloric acid in the stomach.

state

noun

A substance can exist as a solid, a liquid or a gas. This is called its state.

eg At room temperature, the state of water is liquid.

stationary

adjective

When something is stationary it is not moving.

eg When parked, a car is said to be stationary.

🔒 Do not confuse the word 'stationary' with the word 'stationery'. Stationery refers to paper and envelopes, etc.

steam

noun

Steam is the gas or vapour that is produced when water is heated to 100°C and boils.

eg When steam is cooled, it condenses and water is formed.

steel

noun

Steel is an alloy of iron with a small percentage of carbon.

eg The hardness of steel depends upon the percentage of carbon present. Stainless steel is formed when other metals such as chromium and nickel are added.

sterilise

verb

To sterilise is to make something germ-free or to make someone or something incapable of producing offspring.

eg It is necessary to sterilise a baby's feeding bottle before using it.

stigma

noun

The stigma is the sticky surface at the tip of the style of a flowering plant.

eg The stigma receives the pollen.

🔒 The word 'stigma' comes from the Greek word meaning 'brand'.

➡ **petal**

stoma (plural: stomata)

noun

A stoma is a small opening found mostly on the underside of a leaf, where carbon dioxide enters and oxygen and water escape.

eg A stoma can open and close. At night stomata are closed to prevent too much water escaping.

➡ **leaf**

stomach

noun

The stomach is part of the digestive system where the food is temporarily stored and partially digested.

eg Conditions are acidic inside the stomach.

➡ **digestive system**

streamlined

adjective

A body shape that offers the minimum resistance to a gas or liquid is said to be streamlined.

eg Cars are designed using a wind tunnel to ensure the shape is streamlined.

➡ resistance

strength of evidence

noun phrase   ? ? ?

The strength of evidence is an assessment of how reliable investigation results are. Just like a court case, there needs to be enough evidence to come to a sound conclusion.

eg The strength of evidence is such that a conclusion can be safely made.

style

noun

The style is part of a flower. It connects the stigma to the ovary.

eg In some flowers, the style is long and in others it is short. The function of the style is to place the stigma in the best place to receive the pollen.

➡ petal

sufficient data

noun phrase   ? ? ?

To have sufficient data means to have enough information. When carrying out an investigation it is important to collect sufficient data to ensure your conclusions are reliable.

eg Remember that once you have finished an investigation, you cannot go back. Make sure you collect sufficient data as you work.

sugar

noun

A sugar is a carbohydrate. Glucose and fructose are simple sugars (called single sugars or monosaccharides) and sucrose is a more complex sugar, or a disaccharide.

eg Monosaccharides have a formula $C_6H_{12}O_6$ and disaccharides have a formula $C_{12}H_{22}O_{11}$. Both these are sugars.

sulphate (or sulfate)

noun

A sulphate is formed when a metal, metal oxide or metal carbonate reacts with sulphuric acid.

eg Sodium sulphate is formed when sulphuric acid reacts with sodium hydroxide:

sodium hydroxide + sulphuric acid → sodium sulphate + water

surface area

noun

The total area of a surface is called its surface area.

eg A cube with sides of 2 cm has a surface area of 24 cm². This cube can be cut into eight cubes with sides of 1 cm. These cubes have a combined surface area of 48 cm².

suspension

noun

A suspension is a mixture containing solid particles dispersed in a liquid or a gas. The mixture will settle on standing.

eg A bottle of milk of magnesia consists of a suspension of magnesium hydroxide in water.

sustainable development

noun

Sustainable development allows changes and improvements to be made in people's lives but without damaging the environment.

eg Building houses using wood from managed forests is an example of sustainable development.

A B C D E F G H I J K L M N O P Q R **S** T U V W X Y Z

symbol

noun

A symbol is something that represents or stands for something else. In science, each element is represented by a chemical symbol. This is one or two letters with the first letter a capital letter.

eg The symbols for hydrogen, helium and mercury are H, He, Hg, respectively. What is the symbol for oxygen?

symbol equation

noun

A chemical reaction can be summarised by a symbol equation.

eg The reaction of magnesium with sulphuric acid can be summarised by a symbol equation:

$$Mg + H_2SO_4 \rightarrow MgSO_4 + H_2$$

➡ reaction, word equation

synthesis (verb: synthesise)

noun

1 Synthesis is a reaction during which a compound is formed from its constituent elements.

eg The synthesis of water takes place when hydrogen burns in oxygen.

2 The process of putting separate parts together to form a whole is called synthesis.

eg The scientist brought together evidence from various sources for the synthesis of his argument.

The word 'synthesis' comes from two Greek words: *syn*, meaning 'together', and *thesis*, meaning 'placing'.

Tt

taxonomic group

noun

All living organisms are classified into groups according to their shared features. These groups are called taxonomic groups. The largest taxonomic group, which encompasses all living things, is the kingdom. The smallest taxonomic group is the variety.

eg You can remember the order of the taxonomic groups from the largest to the smallest by learning this sentence:

kindly **p**lace **c**over **o**n **f**resh **g**reen **s**pring **v**egetables

kingdom → **p**hylum → **c**lass → **o**rder– **f**amily → **g**enus → **s**pecies → **v**ariety

teeth
➡ tooth

temperature

noun ? ? ?

Temperature is the measure of the hotness or coldness of a body. It is measured on a temperature scale.

eg The temperature is often measured using a thermometer or a temperature probe attached to a computer.

i In science, we usually measure the temperature in degrees Celsius. Before 1948 this temperature system was called Centigrade. On this scale, water freezes at 0°C and boils at 100°C. The lowest possible temperature under any conditions is called absolute zero and is –273°C.
➡ heat

tension

noun

When a body is subjected to stress, a reaction force is set up to counter it. This is called tension.

eg When a string is pulled at its ends, tension is set up equal in mass but in the opposite direction to the pull.

test

noun ? ? ?

A test is a critical examination of qualities or abilities.

eg The test for oxygen is to put a glowing splint into the gas. If the gas is oxygen, the glowing splint will relight.

testis (plural: testes)

noun

The testis is an organ that produces sperm in male animals. The animal usually has a pair of testes.

eg In vertebrates the testes are internal, but in most mammals the pair of testes descend from outside the body in a sac called the scrotum.
➡ reproductive systems

theory (plural: theories)

noun ? ? ?

A theory is a series of ideas or principles that attempt to explain observations and that can be supported by some experimentation.

eg Einstein developed the theory of relativity.
➡ hypothesis, scientific method

thermometer

noun

A thermometer is an instrument used to measure temperature. It is usually a sealed tube filled with mercury or alcohol. As the temperature rises, the liquid inside the tube expands and moves further up the scale.

(eg) The diagram shows a thermometer and a thermometer scale.

What is the temperature shown on the thermometer?

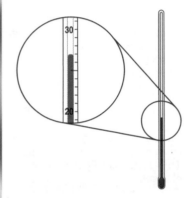

thorax

noun

The part of the human body and the bodies of other vertebrates that is behind the ribs is called the thorax. It also refers to the part of an insect between the head and the body.

(eg) The thorax is often called the chest.
➡ abdomen, diaphragm

time ???

noun

Time is an occasion or period measured in seconds, minutes and hours. The SI unit of time is the second.

(eg) Sam measured a time as one minute and 20 seconds. She wrote down 1.20 m. Why is this incorrect?

time-lapse photography

noun ???

A process that occurs very slowly can be followed by taking a series of photographs at regular intervals. Viewing the photographs in order gives a speeded-up impression of what happens. This is called time-lapse photography.

(eg) Time-lapse photography can be used to show the stages in the growth of a seedling.

tissue

noun

Tissue is a collection of cells that perform a similar function.

(eg) Nerve and muscle are two types of tissue in an animal.

tobacco

noun

A tobacco plant is any plant from the genus *Nicotiana*. It is native to tropical parts of America. It contains a poisonous substance called nicotine.

(eg) Tobacco is used to make cigarettes, cigars and, in a powdered form, snuff.

tonne ???

noun

A tonne is a unit of measurement of mass on the metric system, sometimes called a metric ton.

(eg) A tonne is 1000 kg.

tooth (plural: teeth)

noun

A tooth is one of a set of hard, bone-like structures in the mouth of vertebrates, used for biting and chewing food. Teeth are also used for attack and defence.

(eg) Each tooth consists of an enamel coat, dentine and an inner pulp cavity containing blood vessels and nerves.

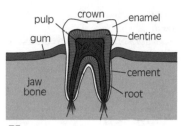

pulp · crown · enamel · gum · dentine · jaw bone · cement · root

ℹ In humans, the first set of teeth consists of 20 milk teeth. These are then replaced and an adult has 32 teeth.

toxin

noun

Any chemical that can damage a living body is called a toxin.

eg In vertebrates, toxins are broken down by enzymes, usually in the liver.

ℹ The word 'toxin' comes from the Latin word *toxicum*, which means 'poison'.

trace

noun

A tiny amount of something, almost too small to detect, is called a trace.

eg Fertilisers contain very small quantities of some elements and these are called trace elements.

trachea (*tra-kee-ya*)

noun

The trachea or windpipe is the tube that conducts air from the larynx in the throat to the two bronchi that lead into the lungs.

eg Around the trachea there are rings of cartilage that prevent the trachea collapsing with changes in pressure.

➡ **alveolus, bronchiole, bronchus, lung**

transect

noun

A transect is the name for a strip of land selected to monitor the changes in composition of vegetation in a selected area.

eg Ecologists will use a transect to study changes in the vegetation of a particular area.

translucent (*trans-loo-sunt*)

adjective

A material is said to be translucent when it allows light to pass through but it is not transparent.

eg Some plastics are translucent, which means that light diffuses through them but you cannot see through them.

transmission

noun

Transmission is the act of passing something on.

eg The transmission of energy from the engine to the wheels of a car enables the car to move.

ℹ The word 'transmission' comes from the Latin word *transmissio*, which means 'sending across'.

transparent

adjective

A material is said to be transparent when it can be seen through.

eg A glass block is transparent.

ℹ The word 'transparent' comes from the Latin word *parere*, which means 'to appear'.

➡ **opaque**

tri- ? ? ?

prefix

The prefix 'tri-' means 'three'.

eg How many sides does a triangle have?

trial ? ? ?

noun

The act of trying or proving something through a test or experiment is called a trial.

eg The trial showed that children using fluoride toothpaste were less likely to need dental fillings than children who did not.

trial measurements ? ? ?
noun

In an investigation, preliminary measurements are taken to ensure that the equipment is all working and that the conditions are correct. These early results, called trial measurements, will not be used in the investigation results proper.

eg Trial measurements should tell you whether the conditions you are using are going to give you the sort of readings you want.

trial run ? ? ?
noun

A trial run is a preliminary run through of the whole investigation.

eg A trial run might be a good idea, particularly if you are uncertain of the outcome.

trophic level
noun

The trophic level is the position a species occupies in a food chain.

eg Plants are at the lowest trophic level – they are primary producers. Herbivores (primary consumers) eat plants and are at the second trophic level. Carnivores (secondary consumers) are at the third trophic level.

trustworthiness of data
➡ data reliability

tune

verb

To tune something is to adjust it so that it works at its best.

eg A musician needs to tune his violin by making fine adjustments so it makes exactly the correct sounds.

tuning fork
noun

A tuning fork is a small, metal device consisting of a stem with two prongs. When the prongs are made to vibrate they give out a particular musical note.

eg A tuning fork is used for tuning musical instruments.

The diagram shows a tuning fork.

i The tuning fork was first invented in the 18th century by John Shore, a trumpeter.

turning effect
➡ moment

ultraviolet [uv]

adjective

Ultraviolet radiation is electromagnetic radiation with wavelengths in the range 4 to 400 nm. It is abbreviated to uv.

eg Ultraviolet radiation is between violet light and X-rays.

umbilical cord

noun

The umbilical cord is a long, tube-like organ which links a fetus to the placenta.

eg The fetus receives nutrients through the umbilical cord.

universal indicator

noun

Universal indicator is a mixture of different indicators that can be used to find the pH value of a substance.

eg The diagram on the right shows the colour of a universal indicator in solutions of different pH values.
➡ indicator

upthrust

noun

When an object is immersed or partially immersed in a liquid or a gas, there is an upward force or upthrust (push) which acts against the weight of the object.

eg According to Archimedes' principle, the upthrust is equal to the weight of the liquid or gas displaced by the object.

using secondary sources

verb phrase   ? ? ?

In practical assessments you should use scientific knowledge and understanding to support your planning. This might mean using secondary sources, such as books, CD-Roms and the Internet.

eg If you are using secondary sources, it is important to write down where you found them and who wrote them so that they can be checked.

uterus

noun

The uterus is the part of the female reproductive system where the fetus develops.

eg The lining of the uterus changes during the menstrual cycle.
➡ menstruation, reproductive systems

pH		Examples
1	STRONG ACIDS	car battery acid, mineral acids
2		
3		
4		lemon juice, vinegar, ethanoic acid
5	WEAK ACIDS	
6		soda water, carbonic acid
7	NEUTRAL	water, salt, ethanol
8	WEAK ALKALIS	soap, baking powder, sodium hydrogencarbonate
9		
10		ammonia solution
11	STRONG ALKALIS	washing soda
12		oven cleaner
13		sodium and potassium hydroxides
14		

Vv

vaccination

noun

A vaccination is the introduction of modified viruses or bacteria into the body so that the body will produce the correct antibody to make it immune to the disease.

eg A vaccination can be by mouth, by a scratch on the skin or, most often, by injection with a hypodermic syringe.

vacuole (*vack-you-ole*)

noun

A vacuole is a fluid-filled cavity inside a cell. It can be a reservoir for fluids that the cell will later secrete or it can be filled with waste products or nutrients that the cell needs to store.

eg Plant cells have large vacuoles for storage.

vagina

noun

The passage from the uterus to an external opening in a female animal is called the vagina.

eg During intercourse, the vagina receives the penis. It also serves as the birth canal down which the fetus travels during delivery.

➡ **reproductive systems**

validity of conclusion

noun phrase ???

After reaching a conclusion, it is important to check how likely to be correct it is by looking at other data. When the conclusion has been supported by other evidence, it can be said to have validity of conclusion.

eg In an experiment, Tim concluded that when a substance burns there is a change in mass. He could check the validity of conclusion of this by checking the results of other people.

vapour

noun

A vapour is a gas that can be condensed just by increasing pressure alone without cooling.

eg When water vapour condenses, water is formed.

🄷 The word 'vapour' comes from the Latin word *vapor*, which means 'steam'.

variable

noun ???

A variable is a property under investigation.

eg All variables, apart from the one being studied, should be kept constant.

variable resistor

noun

A variable resistor is a piece of equipment used to alter the current in an electrical circuit.

eg The diagram shows one type of variable resistor.

➡ **circuit symbol**

variation (variety)

noun

Variation is the difference between individuals of the same species. The causes of variation may be genetic (e.g. colour of hair) or environmental (e.g. earring or not), or a combination of the two.

eg Interesting studies about variation have been carried out with twins brought up in different environments.

vegetation cover

noun

Vegetation cover is the range and variety of different plants covering an area of ground.

eg Year 9 used a line transect to study the vegetation cover in their school grounds.

vein

noun

A blood vessel that carries blood back to the heart is called a vein. The term is also used for any tubes that strengthen living tissue and supply nutrients, e.g. leaf veins.

eg There are valves in veins to stop the blood flowing backwards.

➡ **artery, capillary, heart**

velocity [m/s or km/h]

noun

Velocity is the speed of a body in a particular direction or its change in displacement per second.

eg The units of velocity are metres per second or kilometres per hour.

ventilation

noun

Ventilation is the process whereby the blood takes up oxygen in the lungs.

eg Ventilation takes place when air is inhaled into the lungs.

i The word 'ventilation' comes from the Latin word *ventilare*, which means 'to fan'.

vertebrate

noun

A vertebrate is any animal with a backbone.

eg Mammals, birds, reptiles, amphibians and fish are all examples of a vertebrate. There are 41 000 species of vertebrates.

i The word 'vertebrate' comes from the Latin word *vertebra*, which means 'spinal joint'.

➡ **invertebrate**

vibration (verb: vibrate)

noun

A rapid movement backwards and forwards is called a vibration.

eg Sound is produced when vibration of an object takes place.

i The word 'vibration' comes from the Latin word *vibrare*, which means 'to tremble'.

villus (plural: villi)

noun

A villus is a small finger-like projection, lining and extending into the interior of the small intestine. A large number of villi means there is a very large surface area for absorption of nutrients through the gut wall.

eg Chemicals such as glucose and amino acids pass into each villus and are carried away in the bloodstream.

The diagram shows part of the surface of the small intestine.

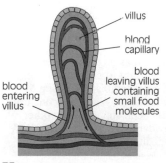

villus

blood capillary

blood leaving villus containing small food molecules

blood entering villus

i The word 'villus' comes from a Latin word meaning 'shaggy hair'.

virus (plural: viruses)

noun

A virus is a kind of micro-organism that can cause disease. Viruses cannot reproduce themselves and can only reproduce inside other cells. When the new viruses burst out of a host cell, the host cell is damaged. Viruses are not destroyed by antibiotics.

eg The diagram below shows a magnified view of a virus.

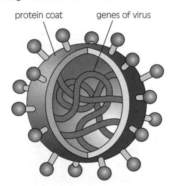

protein coat   genes of virus

The diagram below shows how viruses reproduce in other cells.

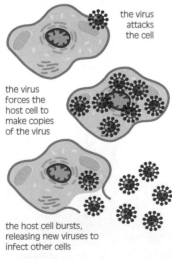

the virus attacks the cell

the virus forces the host cell to make copies of the virus

the host cell bursts, releasing new viruses to infect other cells

➡ bacteria

vitamin

noun

A vitamin is a chemical required in small amounts by the body to control vital processes. Vitamins can be water soluble (B and C) or fat soluble (A, D, E and K).

eg The table gives the sources and uses of some vitamins.

vitamin	source	use
A	green vegetables, butter, egg yolks, fish oils	healthy skin, night blindness
B complex	yeast extract, liver, wholemeal bread	various, particularly respiration
C	citrus fruits, blackcurrants, vegetables	healthy skin, resistance to colds
D	butter, egg yolks, made in the skin	helps make bones

volcanic ash

noun

Volcanic ash consists of rock, minerals and volcanic glass fragments smaller than 2 mm in diameter. It escapes from a volcano during an eruption. Volcanic ash is not the same as the soft fluffy ash that results from burning wood, leaves or paper. It is hard, does not dissolve in water and can be extremely small – ash particles less than 0.025 mm (1/1000th of an inch) in diameter are common.

eg Volcanic ash can be a serious hazard to aircraft even thousands of miles from an eruption. Airborne ash can diminish visibility, damage flight control systems and cause jet engines to fail.

volcano (plural: volcanoes)
noun

The vent through which magma is or has been forced onto the surface of the Earth is called a volcano. Usually the vent is built up into a cone of solidified lava.

eg The picture shows a volcano.

i The word 'volcano' comes from the Latin *Vulcanus,* or *Vulcan*, the Roman god of fire.

voltage
➡ potential difference

voltmeter
noun

A voltmeter is an instrument used to measure the voltage across a component.

eg The diagram shows how the voltage across a lamp can be measured using a voltmeter.

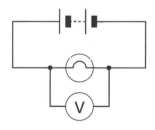

volume
noun

The volume of something is the amount of space it takes up in three dimensions.

eg The volume of the milk in a carton was 1 dm³. This is also called one litre.

What are the names of three pieces of apparatus we use in the laboratory to measure volume?

washing soda

noun

Washing soda is sodium carbonate crystals, $Na_2CO_3 \cdot 10H_2O$.

eg Washing soda is used to soften water in hard water areas.

water

noun

Water is a colourless, odourless liquid with a formula H_2O. It is essential for life.

Water is an excellent solvent for a wide range of substances. Pure water freezes at 0°C and boils at 100°C. Water has its maximum density at 4°C.

eg In a pond, ice forms on the surface as ice is less dense than water. The ice forms an insulating layer which stops the water beneath from freezing.

water cycle

noun

There is a natural cycle of water within the biosphere. This is called the water cycle.

(1) Water in rivers, lakes and the sea evaporates into the atmosphere. Water is also lost by transpiration of plants.

(2) The water in the atmosphere forms clouds.

(3) When the clouds cool, water vapour condenses and falls as rain or snow.

(4) The water drains into rivers, lakes and the sea.

eg The diagram shows the water cycle.

➡ biosphere

wave

noun

A wave is a disturbance travelling through a medium.

eg There are two types of wave:

● longitudinal waves (e.g. sound) consist of a series of compressions and rarefactions

● transverse waves (e.g. water waves, and light and radio waves) involve vibrations at right angles to the direction of wave travel.

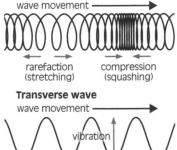

Longitudinal wave

wave movement ⟶

rarefaction (stretching)   compression (squashing)

Transverse wave

wave movement ⟶

vibration

weathering

noun

Weathering is the breaking down of exposed rocks by the action of rain, frost or wind, etc. There are two main types of weathering – physical weathering and chemical weathering. Weathering is different from erosion because there is no

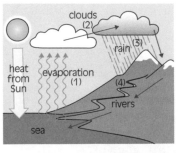

clouds (2)

rain (3)

heat from Sun

evaporation (1)

rivers

sea

movement or transportation of
the fragments of rock.

 Physical weathering takes place
when water gets into a crack in a rock.
As it freezes, the water expands and
forces the rock apart. This is shown in
the diagram

Water in crack
in rock

Ice forms –
expansion forces
rock apart

Chemical weathering is brought about
by a chemical change such as occurs
when rainwater containing carbonic acid
comes into contact with carbonate rocks.
➡ erosion

weedkiller

noun

Weedkiller is another name for
a herbicide. A weedkiller is a
chemical that kills some or all
plants.

 Sodium chlorate is a weedkiller that
can be used to kill weeds on paths as it
kills all plants.
➡ herbicide

weight

noun

Weight is a force exerted on an
object by gravitational attraction.
The weight of an object is
determined by its mass (m) and
Earth's gravitational attraction (g):

$$W = mg$$

The unit of weight is the Newton
as weight is a force.

 The Earth's gravitational field strength
is 10 N/kg. An object of mass 12 kg has a
weight of 120 N. On the Moon the
gravitational field strength is only one-
sixth of that on Earth. What would the 12
kg object weigh on the Moon?

 The word 'weight' comes from
an Anglo-Saxon word, *wiht*.
➡ mass

white blood cell

noun

White blood cells are part of
the blood and are part of the
body's defences. They:
● engulf and destroy invading
  micro-organisms
● produce antibodies
● produce antitoxins that
  counteract toxins produced
  by microbes.

 This diagram shows how a white
blood cell engulfs and destroys a
micro-organism.

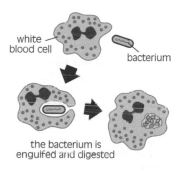

white
blood cell

bacterium

the bacterium is
engulfed and digested

 Human blood contains about
11 000 white blood cells and 500 red
blood cells in each cubic centimetre.

word equation

noun

A summary of a chemical
reaction in words is called a
word equation. The substances
that react (reactants) are on the
left hand side and the substances
produced (products) are on the
right-hand side.

 Carbon burns in oxygen to
form carbon dioxide. This can be
represented by:

carbon + oxygen → carbon dioxide
  (reactants)   (products)

Write a word equation for the reaction
that forms water when hydrogen
burns in oxygen.
➡ reaction, symbol equation

xylem (*zy-lem*)

noun

Xylem is a type of tissue found in some plants. Its main function is to transport water from the roots to other parts of the plant.

eg The diagram shows a cross-section of a stem of a plant in which the xylem can be seen.

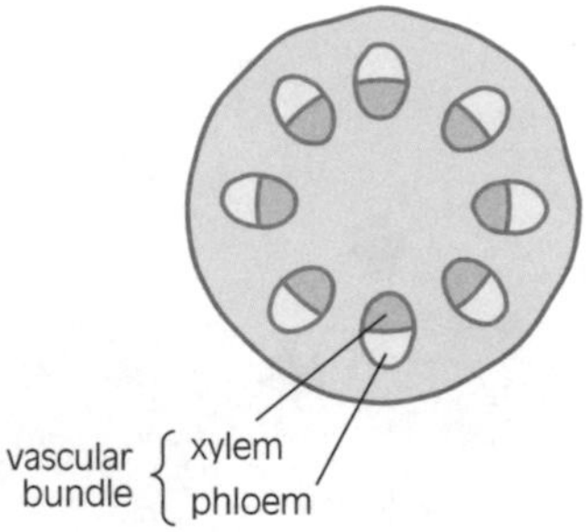

vascular bundle { xylem
phloem

➡ **phloem**

Yy

yeast

noun

Yeast is a single-celled fungus. In a sugar solution, the yeast cells multiply and produce alcohol (ethanol) and carbon dioxide by anaerobic respiration (fermentation).

Yeast is used in baking, brewing and wine making.

yield

noun

The yield is the total amount of something produced by an animal or plant.

New varieties of wheat produce a higher yield of grain than older varieties.

The word 'yield' comes from the Anglo-Saxon word *gieldan*, which means 'to pay'.

Z z Zz Zz Iı

zinc [Zn]

noun

Zinc is a brittle bluish-white metal, widely used in alloys (e.g. brass, copper and zinc) and for coating steel to prevent rusting (galvanising).

eg Zinc is extracted from zinc blende, its ore.

i Zinc gets its name from the German word *zink*.

zoology

noun

Zoology is the branch of Biology concerned with the study of animals.

eg A scientist who studies zoology is called a zoologist.

zygote

noun

The cell formed after fertilisation by the joining of a male gamete (sperm) and a female gamete (ovum) is called a zygote.

eg In a human zygote, the gametes contain 23 chromosomes and the zygote contains 46 chromosomes.

i The word 'zygote' comes from the Greek word *zygon*, which means 'yoke'.

The Periodic Table

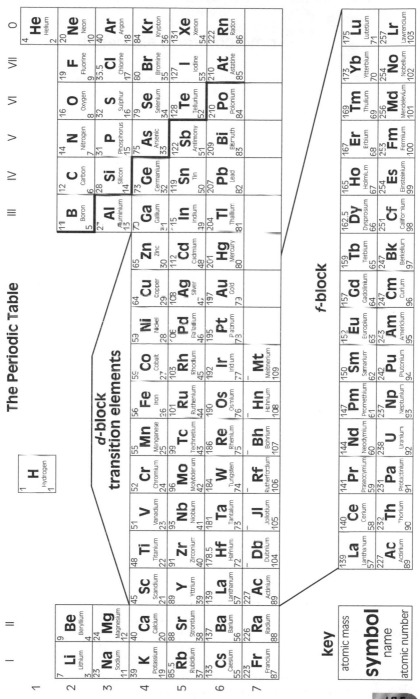

key

| atomic mass |
| **symbol** |
| name |
| atomic number |

d-block transition elements

f-block

Some common elements

Element	Symbol	Metal or non-metal	Melting point (°C)	Boiling point (°C)	Density (g/cm³)	Date of discovery
Hydrogen	H	non-metal	−259	−253	0.00008	1766
Helium	He	non-metal	−270	−269	0.00017	1868
Lithium	Li	metal	180	1330	0.53	1817
Carbon	C	non-metal		4200	2.2	*
Nitrogen	N	non-metal	−210	−196	0.00117	1772
Oxygen	O	non-metal	−219	−183	0.00132	1774
Fluorine	F	non-metal	−220	−188	0.0016	1886
Neon	Ne	non-metal	−249	−246	0.0008	1898
Sodium	Na	metal	98	890	0.97	1807
Magnesium	Mg	metal	650	1110	1.7	1808
Aluminium	Al	metal	660	2060	2.7	1825
Silicon	Si	non-metal	1410	2700	2.4	1825
Phosphorus	P	non-metal	44	280	1.8	1669
Sulphur	S	non-metal	119	444	2.1	*
Chlorine	Cl	non-metal	−101	−35	0.003	1774
Argon	Ar	non-metal	−189	−189	0.0017	1894
Potassium	K	metal	64	760	0.86	1807
Calcium	Ca	metal	850	1440	1.6	1808
Manganese	Mn	metal	1250	2000	7.4	1774
Iron	Fe	metal	1540	3000	7.9	*
Copper	Cu	metal	1080	2500	8.9	*
Zinc	Zn	metal	419	910	7.1	17th Century
Bromine	Br	non-metal	−7	58	3.1	1826
Krypton	Kr	non-metal	−157	−153	0.0035	1898
Silver	Ag	metal	961	2200	10.5	*
Iodine	I	non-metal	114	183	4.9	1811
Barium	Ba	metal	710	1600	3.5	1805
Gold	Au	metal	1060	2700	19.3	*
Mercury	Hg	metal	−39	357	13.6	*
Lead	Pb	metal	327	1744	11.3	*

* These elements have been known for thousands of years.

Important scientific relationships

$$\text{pressure} = \frac{\text{force}}{\text{area}} \qquad \text{force} = \text{pressure} \times \text{area} \qquad \text{area} = \frac{\text{force}}{\text{pressure}}$$

$$\text{resistance} = \frac{\text{voltage}}{\text{current}} \qquad \text{voltage} = \text{resistance} \times \text{current} \qquad \text{current} = \frac{\text{voltage}}{\text{resistance}}$$

$$\text{density} = \frac{\text{mass}}{\text{volume}} \qquad \text{mass} = \text{density} \times \text{volume} \qquad \text{volume} = \frac{\text{mass}}{\text{density}}$$

$$\text{speed} = \frac{\text{distance}}{\text{time}} \qquad \text{distance} = \text{speed} \times \text{time} \qquad \text{time} = \frac{\text{distance}}{\text{speed}}$$

$$\text{work done} = \text{force} \times \text{distance}$$

$$\text{kinetic energy} = \tfrac{1}{2} \times \text{mass} \times (\text{speed})^2$$

$$\text{weight} = \text{mass} \times \text{gravitational field strength}$$

$$\text{average acceleration} = \frac{\text{change in velocity}}{\text{time}}$$

$$\text{moment} = \text{force} \times \text{perpendicular distance to the pivot}$$

Some common compounds and their formulas

Name	Formula
carbon dioxide	CO_2
water	H_2O
sulphuric acid	H_2SO_4
hydrochloric acid	HCl
nitric acid	HNO_3
sodium hydroxide	$NaOH$
potassium hydroxide	KOH
calcium hydroxide	$Ca(OH)_2$
sodium hydrogencarbonate (bicarbonate of soda)	$NaHCO_3$
carbonic acid	H_2CO_3
manganese(IV) oxide	MnO_2
sodium chloride (salt)	$NaCl$
sodium sulphate	Na_2SO_4
sodium nitrate	$NaNO_3$
iron(III) oxide	Fe_2O_3
aluminium oxide	Al_2O_3
magnesium oxide	MgO
calcium carbonate	$CaCO_3$
methane	CH_4
sucrose (sugar)	$C_{12}H_{22}O_{11}$
glucose	$C_6H_{12}O_6$
fructose	$C_6H_{12}O_6$
ethanol (alcohol)	C_2H_5OH